CONDITIONED TASTE AVERSION

OXFORD PSYCHOLOGY SERIES

Editors

1. *The neuropsychology of anxiety: an enquiry into the functions of septohippocampal system*
   Jeffrey A. Gray
2. *Elements of episodic memory*
   Endel Tulving
3. *Conditioning and associative learning*
   Nicholas J. Mackintosh
4. *Visual masking: an integrative approach*
   Bruno G. Breitmeyer
5. *The musical mind: the cognitive psychology of music*
   John Sloboda
6. *Elements of psychophysical theory*
   Jean-Claude Falmagne
7. *Animal intelligence*
   Edited by Lawrence Weiskrantz
8. *Response times: their role in inferring elementary mental organization*
   R. Duncan Luce
9. *Mental representations: a dual coding approach*
   Allan Paivio
10. *Memory, imprinting, and the brain*
    Gabriel Horn
11. *Working memory*
    Alan Baddeley
12. *Blindsight: a case study and implications*
    Lawrence Weiskrantz
13. *Profile analysis*
    D. M. Green
14. *Spatial vision*
    R. L. DeValois and K. K. DeValois
15. *The neural and behavioural organization of goal-directed movements*
    Marc Jeannerod
16. *Visual pattern analysers*
    Norma V. Graham
17. *Cognitive foundations of musical pitch analysis*
    C. L. Krumhansl

18. *Perceptual and associative learning*
G. Hall
19. *Implicit learning and tacit knowledge*
A. S. Reber
20. *Neuropsychology of communicative behaviour*
D. Kimura
21. *The frontal lobes and voluntary action*
R. E. Passingham
22. *Classification and cognition*
W. Estes
23. *Vowel perception and production*
B. D. Rosner and J. B. Pickering
24. *Visual stress*
A. Wilkins
25. *Electrophysiology of mind*
Edited by M. D. Rugg and M. G. H. Coles
26. *Attention and memory: an integrated framework*
N. Cowan
27. *The visual brain in action*
A. D. Milner and M. A. Goodale
28. *Perceptual consequences of cochlear damage*
B. J. C. Moore
29. *Binocular vision and stereopsis*
Ian P. Howard and Brian J. Rogers
30. *The measurement of sensation*
Donald Laming
31. *Conditioned taste aversion: memory of a special kind*
Jan Bures, Federico Bermúdez-Rattoni, and Takashi Yamamoto

Conditioned Taste Aversion

Memory of a Special Kind

Jan Bures
Institute of Physiology,
Academy of Sciences of the Czech Republic, Prague

Federico Bermúdez-Rattoni
Department of Cellular Physiology,
National University of Mexico

and

Takashi Yamamoto
Faculty of Human Sciences,
Osaka University, Osaka, Japan

OXFORD PSYCHOLOGY SERIES
No. 31

OXFORD NEW YORK TOKYO
OXFORD UNIVERSITY PRESS
1998

This book has been printed digitally and produced to a standard design in order to ensure its continuing availability

OXFORD
UNIVERSITY PRESS

Great Clarendon Street, Oxford OX2 6DP

Oxford University Press is a department of the University of Oxford. It furthers the University's objective of excellence in research, scholarship, and education by publishing worldwide in

Oxford New York

Auckland Bangkok Buenos Aires Cape Town Chennai Dar es Salaam Delhi Hong Kong Istanbul Karachi Kolkata Kuala Lumpur Madrid Melbourne Mexico City Mumbai Nairobi São Paulo Shanghai Singapore Taipei Tokyo Toronto

with an associated company in Berlin

Oxford is a registered trade mark of Oxford University Press in the UK and in certain other countries

Published in the United States by Oxford University Press Inc., New York

Reprinted 2002

A catalogue record for this book is available from the British Library

Library of Congress Cataloging in Publication Data
Bures, Jan, physician.
Conditioned taste aversion: memory of a special kind/Jan Bures, Federico Bermúdez-Rattoni, and Takashi Yamamoto.
(Oxford Psychology Series; no. 31)
Includes bibliographical references and indexes.
1. Taste. 2. Conditioned response. 3. Animal behaviour.
I. Bermúdez-Rattoni, Federico. II. Yamamoto, Takashi, 1944– . III. Title. IV. Series.
[DNLM: 1. Taste–physiology. 2. Conditioning (Psychology)
3. Memory–physiology. WI 210 B952c 1998]
QP456.B87 1998 573.8'78—dc21 97-51747

ISBN 0-19-852347-5

Preface

Reasons for writing a book can be very different: reviewing a rapidly growing field of research, announcing a new hypothesis, presenting an original method, describing fruitful cooperation of different disciplines, etc. Although all these motives have played some role in the conception of the present volume, our main intention was to call the attention of neuroscientists to the remarkable phenomenon of conditioned taste aversion, the study of which may, in our opinion, substantially advance the present neurobiology of learning and memory, but which is not yet adequately exploited in analytical research.

Conditioned taste aversion (CTA), known for more than a century to zoologists (Poulton 1887) and pest control specialists (Rzoska 1953), was introduced into the laboratory by John García in the 1950s (García *et al.* 1955) and has since then been intensively investigated by physiological psychologists. This research soon revealed that CTA has a number of unusual properties sharply contrasting with the basic tenets of traditional learning theory. Although some CTA peculiarities (selective associability of gustatory CS and visceral US, hour-long CS–US delay, differential sensitivity of subsequent stages of CTA acquisition to disruption) were recognized more than twenty years ago, they failed to stimulate adequate curiosity of neurobiologists who were unperturbed by the disquieting finding that the critical stage of CTA acquisition may proceed under deep anaesthesia incompatible with other forms of learning. This reluctant attitude to an examination of the facts challenging accepted views has been eroded in the past decade by several pivotal discoveries which considerably advanced our understanding of the neural mechanisms of CTA at the system level and opened up new possibilities of their analysis at the molecular and cellular levels. We believe that the robustness of CTA and the advantages it offers to analytical research will soon make it the preferred model in plasticity studies and we hope that our book will help scientists intending to join this multidisciplinary effort.

The content of the book corresponds to the above aim: Chapter 1 gives a brief survey of the history and present state of CTA research, Chapter 2 describes the basic methodology and terminology of behavioural tests, Chapters 3 to 8 review in detail the experimental evidence obtained by irreversible and reversible lesions as well as by pharmacological, electrophysiological, immunoreactivity, and transplantation studies. Chapter 9 points out the pivotal problems of CTA research and attempts to predict its main direction in the near future. We should like to stress that the rather selective coverage reflects our wish to concentrate on issues corresponding to our expertise. We apologize to readers who may find some of our statements inconsistent or even conflicting. We did not attempt to adopt uniform interpretations of experiments differing by design, initial assumptions, and the scientific traditions of three continents. We believe that this attitude reflects the present state of CTA research, which must be widely open to influences coming from other disciplines and maximally tolerant to views generated by different concepts.

We have worked during the past 20 years in the mainstream of CTA research. Our concepts and experimental approaches were deeply influenced by personal contacts with leading scientists and fellow researchers all over the world as well as by the army of anonymous reviewers of our published papers. To all these colleagues we are grateful for valuable suggestions, technical advice and critical comments. We owe special thanks to John Garcia, the father of CTA research, who was the teacher of one of us (F. B.-R.) and has recruited the other two (J. B. and T. Y.) by his lectures and writings. At the same time, we should like to use this opportunity to thank our former and present co-workers, colleagues and friends, whose names appear in the References, for their contributions to the experiments conducted in our laboratories in Prague, Osaka and Mexico City and summarized in the present volume. Finally we thank the staff of the Oxford University Press for patience, pleasant collaboration, and continuous support.

June 1997

J. B.
F. B.-R.
T.Y.

Contents

List of abbreviations xiii

1 **Ethology, physiological psychology, and neurobiology of CTA** (J. Bures) 1
1.1 Comparative and evolutionary aspects 3
1.2 Long CS–US delay 6
1.3 Neurobiological mechanisms 10

2 **The CTA paradigm: terminology, methods, and conventions** (J. Bures) 14
2.1 Methods 14
2.1.1 Single bottle techniques 14
2.1.2 Multiple bottle procedures 15
2.1.3 Taste reactivity testing 17
2.1.4 Licking behaviour 19
2.1.5 CTA acquired without ingestion 21
2.2 CTA-related phenomena 22
2.2.1 Conditioned taste preferences 22
2.2.2 Changing food classification 23
2.2.3 CTA extinction 23
2.2.4 Attenuation of neophobia 24
2.2.5 Sensitization 24
2.2.6 Conditioned inhibition and higher-order conditioning 24
2.2.7 Latent inhibition 24
2.2.8 Overshadowing 25
2.2.9 Blocking 25

3 **Neuroanatomy of CTA: lesions studies** (F. Bermúdez-Rattoni and T. Yamamoto) 28
3.1 Area postrema 28
3.2 Nucleus tractus solitarius 29
3.3 Pontine parabrachial nucleus 29
3.4 Thalamus 32
3.5 Hypothalamus 33
3.6 Globus pallidus 34
3.7 Amygdala 34
3.7.1 The role of the amygdala in CTA 35
3.7.2 Amygdala and taste-potentiated odour aversions 37

3.8 Hippocampus 39
3.9 Entorhinal cortex 40
3.10 Prefrontal cortex 41
3.11 Insular cortex 41
3.11.1 Insular cortex and taste-potentiated odour aversions 44

4 Functional ablation studies of CTA
(J. Bures) 47
4.1 Gross inactivation procedures 47
4.2 Local inactivation procedures 49
4.2.1 Functional decortication by spreading depression 50
4.2.2 Pharmacological blockade of the insular cortex 51
4.2.3 Pharmacological blockade of extracortical centres 53
4.3 Lateralization of the CTA circuits 56

5 Pharmacology of CTA (F. Bermúdez-Rattoni) 60
5.1 The emetic effects of drugs 60
5.2 Toxicity effects in the formation of CTA 63
5.3 Paradoxical effects of drug reinforcers 64
5.4 Endogenous peptides in the formation of CTA 65
5.5 Stress-related peptides in the formation of CTA 67
5.6 The CS–US delay 69
5.7 Taste interaction with other stimuli 69
5.8 The CS–US–FB model 70
5.9 Neuropharmacology of flavour aversions 71

6 Electrophysiology of CTA (T. Yamamoto) 76
6.1 Taste qualities 76
6.2 Central gustatory pathway 79
6.3 Processing of taste information 80
6.4 General visceral information 82
6.5 Alteration of taste responsiveness after CTA 84
6.5.1 NTS 85
6.5.2 PBN 86
6.5.3 Insular cortex 88
6.5.4 Amygdala 88
6.5.5 Hypothalamus 91

7 Functional morphology of CTA
(T. Yamamoto) 92
7.1 c-*fos* expression after LiCl injection 92
7.2 c-*fos* expression after taste stimulation 96
7.3 c-*fos* expression in the PBN after CTA 96

7.4 CS–US association in the PBN 99
7.5 c-*fos* expression in the NTS after CTA 102
7.6 Other methods 104
7.7 Intracellular events 105

8 Transplantation studies (F. Bermúdez-Rattoni) 108
8.1 Restoration of function by fetal brain grafts 108
8.2 Cholinergic mechanisms in functional recovery 112
8.3 The role of nerve growth factor in recovery 114
8.4 Recovery of learning capabilities 116
8.5 Fetal brain grafts and locus of CTA engram 117

9 Paradoxes, projections, and perspectives of CTA research
(F. Bermúdez-Rattoni, J. Bures, and T. Yamamoto) 121
9.1 CTA and consciousness 121
9.2 Limited role of proteosynthesis 122
9.3 CTA produced by rewarding drugs 124
9.4 Molecular and cellular nature of CTA plasticity 126
9.5 Electrical substitutes for natural CS and US 128
9.6 Conditioned immunity 129

References 133

Index 173

List of Abbreviations

ACh	acetylcholine
ACTH	adreno-cortico-tropic hormone
AN	attenuation of neophobia
ANP	across-neuron pattern
AP	area postrema
APV	amino-phosphonovalerate
ARP	across-region pattern
BLA	basolateral nucleus of the amygdala
BNST	bed nucleus of the stria terminalis
CCK	cholecystokinin
CeA	central nucleus of the amygdala
c-FLI	c-fos-like immunoreactivity
ChAT	choline acetyltransferase
cls	central lateral subnucleus of the PBN
cms	central medial subnucleus of the PBN
CNS	central nervous system
CR	conditioned reaction, conditioned response
CREM	cAMP responsive element modulator
CS	conditioned stimulus
CSD	cortical spreading depression
CT	chorda tympani
CTA	conditioned taste aversion
dls	dorsal lateral subnucleus of the PBN
EC	entorhinal cortex
EEG	electroencephalography
els	external lateral subnucleus of the PBN
ems	external medial subnucleus of the PBN
FB	homoeostatic feedback
GABA	gamma-amino-butyric acid
GLTM	gustatory long-term memory
GN	glossopharyngeal nerve
GP	globus pallidus
GSP	greater superficial petrosal nerve
GSTM	gustatory short-term memory
HRP	horseradish peroxidase
IC	insular cortex
IMP	inosine monophosphate
LH	lateral hypothalamic area

LiCl	lithium chloride
LL	labelled-line
LTM	long-term memory
mdPFC	mediodorsal prefrontal cortex
MeA	medial nucleus of the amygdala
MITC	midline and intralaminar thalamic complex
MSG	monosodium glutamate
MSH	melanine-stimulating hormone
NBM	nucleus basalis magnocellularis
NGF	nerve growth factor
NMDA	N-methyl-D-aspartate
NTS	nucleus tractus solitarius
OFN	orbital frontal neocortex
OT	odour-taste
PBN	parabrachial nucleus
PBNlat	lateral part of the PBN
PBNmed	medial part of the PBN
PFC	prefrontal cortex
PKC	protein kinase C
PSI	protein synthesis inhibitor
SL	superior laryngeal nerve
sPFC	sulcal prefrontal cortex
TTX	tetrodotoxin
UR	unconditioned response
US	unconditioned stimulus
VM	ventromedial nuclear complex of the thalamus
VMH	ventromedial hypothalamic nucleus
VPLpc	parvocellular part of the ventral posterolateral nucleus of the thalamus
VPM	ventral posteromedial nucleus of thalamus
VPMpc	parvocellular part of the VPM of the thalamus

1 Ethology, physiological psychology, and neurobiology of CTA

The existence of organisms depends on their capability to maintain the homeostasis of the internal environment in spite of the continuously changing conditions of the outside world. The antientropic effort required to keep the composition and energy balance of the body within the narrow, genetically pre-set limits necessary for its function concentrates on the flow of substances and energy across the organism–environment interface. In fact, this boundary is rather vague because the organism is continuously entered by gases it inhales, by the water it drinks and by the food it consumes. While the gaseous composition of the atmosphere and the chemical properties of water are relatively stable, feeding is a much more complex process which must provide the animal with a sufficient supply of energy rich edibles as well as with indispensable quantities of vitamins, minerals, and essential amino acids. In this perspective, feeding extends the homeostasis of the internal environment of the organism by making an appropriate part of the outside world into a part of the body and by ejecting undesirable substances and wastes from the body.

The vital importance of these processes led to the development of reliable control mechanisms which subserve in higher organisms various forms of feeding behaviour. They consist of the following main components:

(1) Recognition of the bodily needs, caused by low level of energy rich substrates (manifested by hunger) or by lack of an essential substance (manifested by specific hungers, e.g. for sodium or vitamin B).
(2) Identification of the available substances the ingestion of which may satisfy the outstanding need, finding and consuming sufficient quantities of these substances.
(3) Regulatory processes making ingestion directly proportional to the intensity of the need.

The integration of these components into a neurobiological system was first proposed by Stellar (1954) and later elaborated by Rolls (1975), LeMagnen (1985), and Toates (1986). Evolutionary pressures have influenced feeding behaviour similarly as the anatomy and physiology of the animals and have provided these vital function with robust neural and humoral mechanisms forming the basis of motivation and of the learning and memory phenomena indispensable for the identification of edibles, the consummation of which satisfies different needs. A characteristic feature of associations of this type is that they are formed between visceral states (signalled through visceral afferents) and external stimuli (mediated by gustatory, olfactory, visual, and somatosensory afferents).

Although these associations are sometimes described as classical conditioned reactions, they serve the function of instrumental reactions elicited by deviation of the internal state from the pre-set equilibrium. Presentation of a classically conditioned CS produces a conditioned reaction (CR) mimicking the response to the associated US, e.g. salivation anticipating the delivery of edible food or vomiting anticipating the nausea elicited by toxic food. Classical conditioning only applies to situations when the CS–US contingency cannot be changed by the subject. This is true for the CS–US association stage (flavour of the ingested food is associated with subsequent satiation or malaise produced by the ingested food), but once the CS has acquired the properties of an appetitive or aversive signal the animal reacts to its presentation by instrumental reactions increasing or decreasing exposure to the US and to the ensuing internal states. Selection and ingestion of food (characterized by typical exteroceptive stimuli) may either return the internal state back to normal, may leave it unchanged, or may even increase the disequilibrium. From this point of view the change of the internal state is the UR which assumes either the properties of reward (satiation) or of punishment (poisoning).

Successful feeding must be based on efficient feedback between the properties of the ingested edibles and the consequences of their ingestion. The tighter is this feedback, the finer is the regulation of homeostasis around the predetermined set points (glycemia, pH, $[K^+]_e$, $[Ca^{2+}]_e$, vitamins, osmolarity, etc.), and the better is the quality of the resulting adaptive behaviour. It is obvious that identification of edibles relevant to specific physiological needs can only be accomplished by a demanding learning process associating delayed consequences of feeding with the sensory properties of eaten food. The results range from basic discrimination of edibles from non-edibles to formation of lists of foods with specified nutritional values or with possible toxic effects.

Learning of the above type is a form of adaptation to the changing environment (Staddon 1983). According to ethological views (Seligman and Hager 1972) behaviour of animals is particularly well adapted to conditions with high expectancy and critical survival value. Selection for associative predispositions of food with visceral consequences of its ingestion has led to the formation of 'biological preparedness'. According to Seligman (1970, p. 408) 'The organism may be more or less prepared by the evolution of the species to associate a given CS and US or a given response with an outcome. The relative preparedness of an organism for learning about a situation is defined by the amount of input (e.g. number of trials, pairings, bits of information, etc.) which must occur before that output (responses, acts, repertoire, etc.), which is construed as evidence of acquisition, reliably occurs.'

According to the above operational definition organisms are prepared, non-prepared, or even contra-prepared to learn a class of associations. Learning is fast in the prepared and slow in the contra-prepared animals but proceeds at a normal rate in the non-prepared state. Food selection learning belongs to the category of prepared behaviour relating a specific class of cues (taste, odour, or sight of food) with the consequences of its ingestion (neural and humoral signals monitoring the internal milieu of the organism).

1.1 Comparative and evolutionary aspects

Most of the food selection research is done in rats, but the task is so universal that it is encountered at all levels of evolution. Similar forms of food aversion learning are found in vertebrate and invertebrate species, the ancestral lines of which diverged more than 500 million years ago. The predatory mantis feeding on milkweed bugs which had eaten the cardenolide poison contained in the milkweed, stops eating the poisonous bug, vomits, and rapidly learns to reject this prey on sight (Berenbaum and Miliczy 1983). Closely similar avoidance of the cardiac glycosides containing monarch butterflies is observed in the insectivorous blue jays (Brower and Fink 1985). Since the monarchs use aposematic coloration to warn possible predators, the CS is not necessarily the taste but also the visual appearance of the butterfly. Laboratory studies showed that the blue jays trained to eat non-toxic monarch butterflies learn in less than 20 trials to discriminate them from the toxic ones, the ingestion of which elicited vomiting after 10 to 20 min.

It must be pointed out, however, that the US is not necessarily the illness caused by the ingested toxin but already the bad taste of the poisonous butterflies. In this case, the ensuing learning resembles the pecking suppression task extensively studied in newly hatched chicks (Cherkin 1969; Rose 1995) who are easily discouraged to peck at a bright bead coated with a bitter tasting substance. Although the learned rejection of the unpalatable food may prevent poisoning, it was not due to the association of the CS (sight of food) with the visceral consequences of its ingestion but rather with other sensory properties of the same food (taste, texture). Better documented food selection learning in invertebrates was described in the garden slug (Sahley *et al.* 1981) which learns to associate vegetable flavours and poisons.

In these experiments the slugs (*Limax maximus*) exposed to a 20-min pairing of attractive food odour (potato or carrot) with a bitter tasting plant substance (quinidine sulphate) decreased preference for this odour in a two-choice test performed 24 h later. The preference change was specific for the paired odour and did not generalize to odours not paired with quinidine. Since contact with the quinidine solution elicited an immediate aversive reaction characterized by attempted withdrawal, mucus secretion, and writhing, the association proceeded between the olfactory CS and the gustatory or cutaneous US, although visceral effects of the ingested quinidine cannot be excluded.

In an earlier study (Gelperin 1975) aversion to a new highly palatable taste was induced by its association with subsequent 5-min exposure to CO_2. Weaker aversion was produced when the taste of the food was followed by CO_2 exposure 1 h later. No aversion developed when the taste–CO_2 delay was prolonged to 3 h. Although these experiments have the attributes of long-delay learning and appropriate controls excluded the possibility of CO_2-induced non-specific increase of neophobia, hypercapnia is not a very effective US in CTA studies in rats (Ionescu and Buresova 1977*a*; Nachman and Hartley 1975).

Gelperin (1975) failed to elicit CTA to novel food containing an effective dosage of the slug poison metaldehyde which made the snails very sick after some delay. On the other hand, wild slugs show a striking selectivity for the plants they eat. This behaviour probably reflects not only the taste but also the nutritional value and potential toxicity of secondary substances released during digestion of the leaf tissue. Such secondary effects were demonstrated by Delaney and Gelperin (1986) in experiments showing that Limax can learn to avoid tastes associated with a methionine deficient diet. In a naturalistic setting avoiding certain edibles must necessarily increase preference for other items on the list of available edibles. Conditioned food aversion and conditioned food preference are thus two sides of the same coin.

The separation of the external and internal influences of the US is often quite difficult. This was convincingly illustrated by experiments performed in another mollusc, *Aplysia californica*, which can change its response to palatable food when attempts to consume it systematically fail (Schwarz and Susswein 1986). Aplysia was offered pieces of edible algae strapped in a plastic net which prevented effective biting and swallowing and thus made the food inedible. The frequency of food elicited responses gradually decreased until the animal stopped responding to food objects altogether. In a later elaboration of the test (Chiel and Susswein 1993) Aplysia could freely approach seaweed tied into a canvas bag. A force transducer attached to the bag made it possible to record the number, duration, and intensity of individual attempts. All these measures declined during 2.5 h of the first exposure to the inaccessible food and the decline was considerably accelerated during re-exposure on day 2, provided that the first session lasted 2.5 h or more. Shorter first day exposure (1.25 h) was ineffective.

Food selection learning takes into account not only the nutritional value of various edibles but also their accessibility under specific conditions. The above examples illustrate extinction of a positive food signal due to the failure of the food to enter the gut. Although this information can be obtained already by the perioral receptors and does not require obligatory participation of visceral signals, gut denervation prevents Aplysia from learning that the food is inedible (Schwarz and Susswein, 1986). Bilateral section of oesophageal nerves lead to prolonged attempts to ingest the netted food and to absence of savings during re-exposure. Cessation of feeding attempts in the gut-denervated Aplysias was due to sensory adaptation, habituation, or fatigue and did not result in the formation of long-term memories (Schwarz *et al.* 1988). This indicates that the signals transmitted through the oesophageal nerves are indispensable for both positive and negative reinforcement.

Whereas blockade of positive reinforcement after denervation is understandable, interference with negative reinforcement suggests that the absence of signals is not equivalent to the absence of food in the gut. It is conceivable that the presence of food at the entrance to the gut activates a circuit anticipating the transport of food into the gut. Continued afferentation from the gut signalling that the gut remains empty contrasts with the corollary discharge activated by the consummatory attempts. This difference between the expected state and actual state probably generates the negative reinforcement indispensable for learning that the food is inedible. After section of the oesophageal nerves no signals are arriving from the gut and their absence prevents implementation of both negative and positive reinforcement.

In a series of follow-up experiments Schwarz *et al.* (1991) confirmed that denervation of the gut blocks the ability of the Aplysia to learn that food is inedible but does not affect retrieval of memories formed before the denervation. The authors concluded that the signals carried through the oesophageal nerves are necessary during training but not subsequently, i.e. that they are not required for expression of learned changes of feeding behaviour.

Another consequence of gut denervation is impairment of satiation (Kuslansky *et al.* 1987). The denervated Aplysias require a significantly larger volume of ingested food before they stop feeding. Cessation of feeding is in this case due to prolonged sensory stimulation and ensuing receptor adaptation or habituation. Interpretation of the above finding was complicated, however, by the fact that a section of the oesophageal nerves blocked not only the afferent signals from the gut but also caused paralysis of the oesophageal sphincter and resulting leakage of food from the crop back to the buccal cavity (Kupfermann 1974). This confounding factor was eliminated in experiments using distension of the crop by an inflatable balloon. Inflation of the crop to a volume corresponding to 13% of body weight immediately increased the interbite interval from 10 to 50 s. This inhibition of feeding rapidly disappeared after deflation of the balloon. In denervated Aplysias satiation criterion was reached with a twofold larger crop distension than in intact animals. Similarly, the denervated Aplysias require twice as long as intact animals to reach the satiation criterion. The above experiments do not support the assumption that satiation is mediated by chemostimulation of the rostral part of the gut by substances released from the ingested diet. The critical evidence was obtained in experiments demonstrating that Aplysias filled almost to satiation with non-nutritive bulk do not satiate after ingestion of a small amount of nutritive food providing sufficient chemical stimulation of the rostral gut.

The evidence obtained in invertebrates shows that most problems of food selection learning studied in mammals are encountered at a more primitive level of brain organization. The ability to recognize and reject toxic food on sight, by smell or non-consummatory tasting, is a great biological advantage protecting the animal against fatal poisoning. On the other hand, too cautious food selection may reduce the available food resources below an acceptable level. As pointed out by Forkman (1995) animals are surrounded by many types of potential food, some of which are known as poisonous or edible while other are unknown. Eating only the familiar food is safe as long as its supplies are sufficient. Its disappearance may force the animals to test the unknown and possibly toxic food sources because they have no other choice. Hungry animals on a positive energy budget prefer to eat the rather scarce familiar food because this is less risky than the time and energy consuming exploration of alternative food sources. Novel food is more frequently sampled by sated animals because they have no other immediate needs and exploration of alternative food sources provides them with information which may give them a survival advantage in the long-term perspective (Staddon 1983; Stephens and Krebs 1986). Testing new diets or verifying the continued toxicity of the familiar ones is costly because they may have no nutritional value or be poisonous. Exploration may show, however, that the new food is more profitable than the familiar one or that it can serve as a less profitable but available food source when the supplies of the familiar one are exhausted (Forkman 1993). A sated animal can afford to combine foraging with sampling and in this way trade energy for information, an attitude which applies to other instances of decision making.

1.2 Long CS–US delay

A characteristic peculiarity of CTA learning (García *et al.* 1974; Riley and Tuck 1985) is the long delay (1 to 12 h) between the food related CS (taste, odour, colour, texture) and the aversive internal symptoms elicited by the US (the ingested toxic substance). Since the conventional forms of defence conditioning yield optimum results only when the CS precedes the US by 1–2 s (Abrams and Kandel 1988; Macintosh 1974) this feature of CTA was initially accepted with considerable scepticism (Bitterman 1976). One of the alternative explanations offered was that the delayed CS–US association is mediated by sensory after-effects due to lingering tastes or to regurgitation of ingesta which may support indirect temporal contiguity of the taste signal with the visceral symptoms of poisoning. While regurgitation may apply to slowly digested edibles which are present in the stomach hours after ingestion, the aftertaste hypothesis cannot be upheld in the case of rapidly metabolized substances such as sucrose solutions (Revusky 1968) which are not more present in the stomach and cannot, therefore, be regurgitated, at the onset of the several-hour delayed sickness. Furthermore, regurgitation is improbable in rats which cannot vomit and instead consume non-nutritive substances such as clay or kaolin to remove the poison from the intestine (Mitchell *et al.* 1976). The aftertaste view is not compatible with experiments showing that CTA to a particular novel flavour is not disrupted by ingestion of other flavoured solutions in the CS–US interval (Kalat and Rozin 1971).

The indirect temporal contiguity assumption fails completely in the case of conditioned food aversions based on non-gustatory cues, e.g. the colour or shape of food (Wilcoxon *et al.* 1971; Braveman 1975; Mackay 1977; Brower and Fink 1985), temperature of water (Nachman 1970), or on virtual taste produced by electrical stimulation of the tongue (Weijnen 1977). After a heated discussion of these issues in the late 1970s (Revusky 1977), CTA is now generally accepted as a prototype of long delay learning implemented by mechanisms not requiring close temporal contiguity between activities of the CS and US pathways (Chambers 1990; Schafe *et al.* 1995*a*).

While it is certainly true that CTA acquisition does not require close contiguity or even overlap of the CS and US volleys, Schafe *et al.* (1995*a*) have recently reported that such overlap may even weaken the resulting CTA. Rats received infusion of 2 ml of 0.15% saccharin into the oral cavity at the rate of 0.5 ml/min as the CS and intravenous infusion of 7.5 mg/kg apomorphine as the US. Apomorphine infused 10 s after CS onset produced weaker CTA than infusion of the same apomorphine dosage after 15 or 30 min delays. Since the apomorphine effect (lethargy, stereotype movements, prostration) appeared with a latency of 10 s, sickness culminated before termination of the CS. Several earlier studies reported that backward conditioning (application of the US shortly before presentation of the CS) yields weaker CTA than delayed forward conditioning (Nachman 1970; Barker and Smith 1974; Spector *et al.* 1988; Yamamoto *et al.* 1995). With this procedure, CS may partly precede the culmination of the US effect so that weak CTA can be ascribed to low effectiveness of the short CS–US delays. It is conceivable that the CS presentation induces the formation of a gustatory short-term memory (GSTM) and that it is this trace and not the volleys in the CS pathway which associate with the visceral US.

Although GSTM mainly serves the purpose to bridge long CS–US delays, decreased efficiency of CTA learning with overlapping CS and US suggests that the critical associations are not formed by direct interaction of the CS and US activated circuits but require the mediating GSTM buffer even when there is an opportunity for direct convergence of the CS and US volleys in the circuits mediating aversive labelling of the gustatory experience. The GSTM may appear unnecessary when the taste signal overlaps with poisoning, but experiments show that this is not the case. Bures and Buresova (1981) demonstrated that presentation of the taste CS to functionally decorticated rats prevented subsequent association of the gustatory CS with LiCl poisoning. On the other hand, functional decortication induced after presentation of the CS did not interfere with the effect of the US and did not prevent CTA induction. It appears that cerebral cortex is necessary for the formation of the GSTM at a subcortical level where it waits for the arrival of the US.

Figure 1.1 shows a schematic model of the putative processes underlying CTA acquisition.

The subcortical GSTM centre is accessed by the taste signals through the gustatory cortex. Isolated presentation of the taste cue results in formation of the GSTM, i.e. activation of a circuit priming the specific CS input to the centre implementing formation of the gustatory long-term memory (GLTM) and expression of the CTA response. The persisting activity of the priming circuit even after cessation of the CS carries the taste signal through the CS–US delay. Arrival of the US increases the synaptic weight of the GSTM input to the neurons activated by the US which accounts for the GLTM formation. The GLTM neurons cannot be accessed directly by the CS volley but only through the memory buffer (GSTM) the purpose of which is not only to cope with the CS–US delay but also to provide a device for interaction between the temporarily stored neutral signal and other later-appearing relevant signals and for their eventual transformation into interpreted labelled engrams stored in permanent memory. The activation of the GSTM buffer needs cortical participation but its activity persists even after cortical blockade. The fact that CTA acquisition is disrupted by functional decortication even when CS and US are simultaneously presented indicates that the GSTM buffer cannot be bypassed by a direct CS input to neurons implementing GLTM formation.

The fact that CTA is weaker with simultaneously applied than with delayed CS and US is probably due to slow formation of the GSTM which provides a time window determining the efficiency of the delayed CS–US association. The associability is low during the first minutes of the CS presentation when the GSTM activity is still building up, is maximal from 10 to 60 min, and then

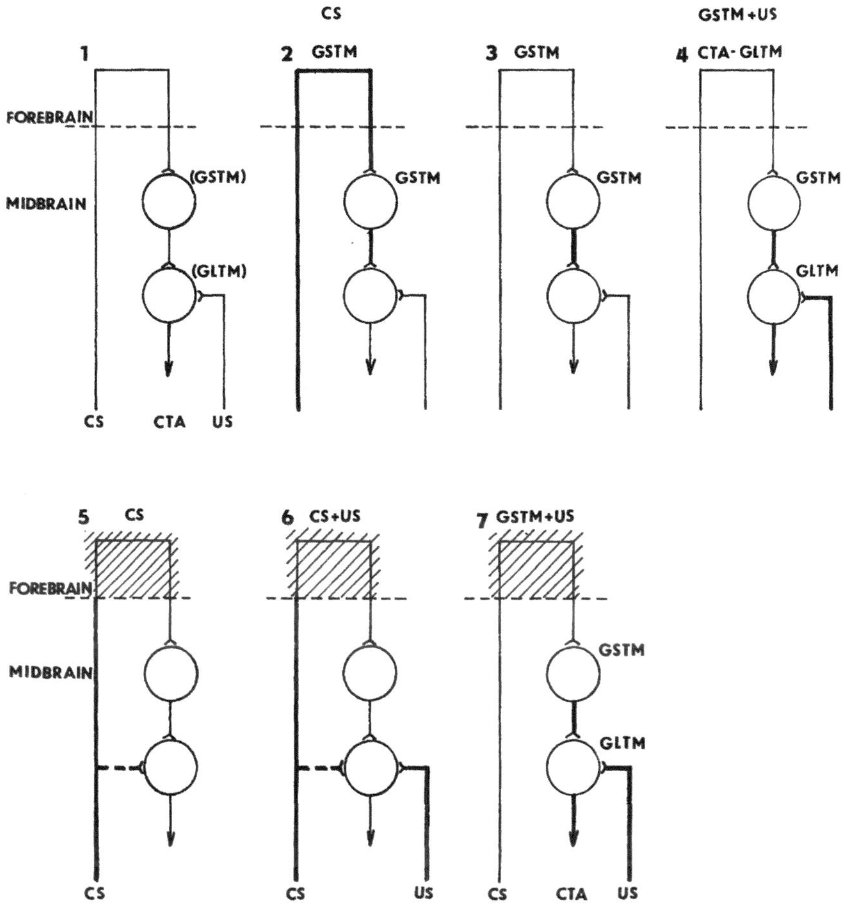

Fig. 1.1 Schematic representation of the neural circuit implementing putative stages of CTA acquisition at the forebrain and midbrain levels. CS, conditioned taste stimulus; US, unconditioned visceral stimulus; GSTM, gustatory short-term memory; GLTM, gustatory long-term memory mediating aversive labelling of the CS and resulting CTA. Heavy lines indicate the part of the circuit that is active during the particular stage of the experiment. Above: CTA acquisition in the intact brain: (1) components of the inactive circuit; (2) GSTM is formed during CS presentation; (3) GSTM persists during the several hour-long CS–US delay; (4) US coinciding with GSTM leads to formation of GLTM for CTA. Below: effects of permanent or temporary elimination of the forebrain (indicated by shading): (5) no CTA can be acquired because no GSTM has been formed during CS presentation; (6) failure of CTA formation by simultaneous presentation of CS and US indicates that GLTM cannot be formed by simple CS–US convergence through putative (dotted line) gustatory inputs; (7) GSTM formed with intact brain persists during forebrain blockade and supports GLTM formation by association with the appropriate US.

slowly decreases as the GSTM decays to zero after 8 to 12 h. The shape of this window corresponds to the probability of the causal relationship between ingestion of the CS and symptoms of poisoning. Transport of ingesta into the intestine, digestion, and resorption take at least a few minutes. It is unlikely that fully fledged toxicosis develops during a few minutes after ingestion and this low probability is expressed by the weak CTA formed. The 10 to 60 min interval corresponds to the most common instances of food aversion learning and yields, therefore, CTA of maximum strength.

Poor efficiency of backward CTA conditioning can also be due to intrinsic taste signals produced during poisoning by regurgitation of stomach juice or bile. These internal gustatory stimuli can be confused with the tastes introduced into the oral cavity from outside. Poisoning may distort perception of simultaneously presented taste stimuli so much that they cannot be detected when presented alone during retrieval testing. The underlying mechanism is related to state dependent learning.

1.3 Neurobiological mechanisms

Various component processes of food selection learning differ according to their long-term and short-term significance. Failure to find a new source of energy rich food or of an essential dietary item does not immediately endanger the life of the animal which can survive prolonged periods of starvation or of deficient diet. Learning about the availability or profitability of food does not require fast decisions but rather careful evaluation of minute visceral consequences of ingestion. This is best documented by changes of neophobia, i.e. cautious intake of small portions of a novel diet followed by several hours of fasting (Domjan 1977). This behaviour makes it possible to separate a particular novel food from the familiar diet and to evaluate its nutritional value more reliably. The taste (and other sensory attributes) of the novel food is stored as GSTM until visceral consequences of ingestion make it possible to evaluate the experience as beneficial. Reduction of hunger, thirst, or symptoms due to deficient diet during several hours after food intake elicits transformation of the GSTM into GLTM with the label 'desirable'. The presence of this record is manifested during subsequent encounters with the same food by attenuation of neophobia (AN), i.e. increased intake of the now already familiar food.

Since AN is the result of uninterrupted integration of visceral signals monitoring the internal state of the organism, it is disrupted by interfering interventions. AN develops slowly: its first signs appear only 1 h after the first presentation of the gustatory stimulus and fully developed AN is observed after 6 h (Nachman and Jones 1974; Green and Parker 1975). As shown by

Buresova and Bures (1980) AN formation can be interrupted by interventions interfering with the underlying integrative activity in the critical post-ingestion interval. Thirsty rats offered a novel flavour (apple juice) demonstrated neophobia by drinking only 3 ml during the first 15-min presentation and AN by increasing apple juice consumption to 7 ml during the second presentation two days later. No AN developed in rats in which pentobarbital anaesthesia (40 mg/kg), deep hypothermia (20°C), or functional decortication by repeated waves of spreading depression was elicited and maintained during the first 2 h after the first apple juice drinking. At least 2 h of uninterrupted monitoring of the consequences of apple juice drinking were required to produce AN. AN was also disrupted by short-lasting interventions applied in the same time window, e.g. by brief ether anaesthesia (5 min) or electroconvulsive shock. This effect indicates that the disrupted AN-forming process cannot be resumed after recovery of brain function. This is understandable, because association of the gustatory experience with incomplete record of its visceral after-effects may lead to incorrect classification. The neural circuits implementing AN include the highest levels of CNS the impairment of which precludes correct interpretation of the gustatory–visceral experience and leads to AN failure.

The situation is quite different in the case of visceral signals of poisoning. The interventions which disrupt AN do not interfere with CTA acquisition, when applied after presentation of the gustatory stimulus and before administration of the poison. This was shown for functional decortication, general anaesthesia, hypothermia, and electroconvulsive shock. A necessary prerequisite of the above experiments, reviewed in Chapter 4, is that the intervention alone elicits no CTA at all or only a weak one (pentobarbital anaesthesia) which does not obscure the effects of the toxin. With the standard CTA procedure, blockade of the higher brain centres and of the arousal mechanism does not interfere with the GSTM storage (Fig. 1.1). Since most of the interventions outlast the CS–US interval and overlap with the maximal US effect, it is obvious that they neither decrease the saliency of the visceral US nor prevent the putative GSTM–US association and formation of the negatively labelled GLTM.

The striking resistance of GSTM to disruption demonstrated in the CTA acquisition experiments suggests that the AN blockade induced by the same interventions is not due to GSTM erasure. It is conceivable that the same GSTM is shared by two independent associative mechanisms implementing its positive or negative labelling. The positive mechanism mediating not only AN but also taste preference learning and the so-called 'medicine effect' (García *et al.* 1974) seems to be more fragile than the more robust CTA circuit. The difference between these two forms of taste signal processing reflects the priorities ascribed by the organism to the short-term and long-term goals: avoiding a fatal poisoning is more important than correcting a long-standing deficiency. This is probably the reason why evolution has equipped animals with brain circuits which remain functional even when the poison has impaired the higher brain functions and induced a comatose state incompatible with conventional forms of learning. In fact, the loss of consciousness may precede the noxious visceral signals which may remain completely below the awareness threshold. From this point of view, CTA can serve as an example of implicit learning, a visceral version of blindsight (Weiskrantz 1986) essential for the animal's coping with the inherent risks of feeding. There is no need to use a similar form of association for somatic signals typically processed by conscious behaving animals which may be in the worst case aroused by the application of the US. The CTA circuit provides automatic recording of the visceral consequences of ingestion which continues even under conditions blocking the conscious record of the information entering the brain e.g. through the ears or other sensory organs. This truly exceptional feature of CTA learning is obviously due to the vital significance of this record which makes it possible to identify in the conscious experience the events which may be causally related to subsequent unconscious nausea and malaise. As pointed out in an earlier publication (Bures *et al.* 1988) the role of CTA can be compared with the role of the flight recorder, an indestructible device on commercial aeroplanes which records all instrument readings and verbal communications between the crew members during the flight. In the case of an accident, the record is available for reconstruction of the preceding events and for the planning of preventive measures.

It must be emphasized that the protective efficiency of the CTA mechanism can only be guaranteed under specific conditions. (1) The association is limited to a few food related events singled out by the feeding habits of the animal (neophobia with the ensuing interruption of eating). (2) The external signal must be processed with intact brain. In particular, the failure to take into account the latter condition may lead to incorrect inferences, e.g. to the statement that CTA can be formed under anaesthesia. This is not the case because most interventions which do not impair the GSTM–US association prevent formation of the GSTM. This is not only due to the fact that they interfere with spontaneous consummatory activity (functional decortication) or with all forms of behaviour (anaesthesia) but applies even to situations when the gustatory stimulus is applied by intraoral infusion (Buresova and Bures 1973; Spector *et al.* 1988; Yamamoto *et al.* 1995) or by systemic application of the taste stimulus by the intravenous (Bradley and Mistretta 1971) or intraperitoneal (Baum *et al.* 1974) routes. When taste stimuli administered in the above way to intact rats are followed by or overlap with poisoning, they elicit the usual, albeit slightly weaker, CTA. When they are applied by forced drinking to functionally decorticated (Buresova and Bures 1973) or hypothermic (Buresova and Bures 1981) or by intraperitoneal injection to anaesthetised (Buresova and Bures 1977, 1979) rats, no GSTM is formed and no CTA is induced by subsequent poisoning (Fig. 1.1). This means that the exteroceptive component of the taste–illness learning is processed in the same way as the exteroceptive cues used in the conventional forms of learning. It is possible that the short-term memory buffer can be filled with other food related exteroceptive signals (appearance, texture, or temperature of food) the relevance of which for the ensuing illness is determined by the act of ingestion. It is conceivable that similar short-term memory files may exist in connection with other visceral states developing as a delayed consequence of an exteroceptively signalled environmental influence transmitted not by ingestion but by respiration or radiation. Smell may perhaps play a role comparable with taste in the case of volatile toxins but no clear evidence for delayed learning of conditioned gas aversion is available, probably because the relatively stable composition of the atmosphere makes reactions of this type less important.

The possibility to establish CTA to vascular taste indicates that ingestion is not an obligatory condition of GSTM formation. This exception is probably only possible with taste stimuli the presence of which implies feeding. Other exteroceptive stimuli can elicit CTA only after association with taste. This is best illustrated by the phenomenon of taste-potentiated odour aversion (Rusiniak *et al.* 1979). Odour alone is a much weaker CS than taste alone in CTA learning but its salience can be highly increased when it overlaps with taste followed by poisoning. When the interval between odour and taste is increased to more than one minute the potentiation is lost. This indicates that the taste-potentiated odour aversion requires close temporal contiguity of the odour and taste components typical for conventional learning and that the long delay learning and resistance to disruption applies only to the taste–visceral link.

2 The CTA paradigm: terminology, methods, and conventions

The capability of animals to select from the available sources of food an energetically adequate diet with properly balanced nutrients, vitamins, minerals, and essential amino acids is a spectacular demonstration of adaptation to uncertain and changing environments (Richter 1943; Staddon 1983; Forkman 1995). The phenomenon is, however, too complex for laboratory analysis which is usually performed under conditions of maximally simplified selection, i.e. the hungry animal is presented with a single novel food when no familiar food is available (García and Koelling 1966). To facilitate quantitative evaluation of ingestive behaviour the diet is usually presented in the form of a liquid the consumption of which can be easily recorded and measured. In the standard experiment, rats are adapted to a 23.75-h water deprivation schedule and trained to drink their daily ration of water during a 15-min stay in the drinking box equipped with one or several drinking pipettes.

2.1 Methods

2.1.1 Single bottle techniques

When using the single bottle procedure the thirsty rat is offered the novel fluid only. The volume consumed depends on the motivation of the animal and on the palatability of the novel taste. The result is compared with the asymptotic intake of water on previous days. When using a moderately unpalatable gustatory stimulus (apple juice, 2% saccharin) reduced fluid intake is a manifestation of neophobia (Domjan 1977). The experiment may be terminated at this stage. If the same fluid is offered under the same conditions on the next day, its consumption can be higher not only by attenuation of neophobia but also by increased thirst motivation. This is the reason why the next presentation of the taste stimulus follows after one or several days of access to water. Increasing consumption of the novel taste during several subsequent presentations can then be considered as due to attenuation of neophobia, i.e. to a reduced fear of the flavour the consumption of which was not followed by adverse consequences but rather by satiation of thirst.

The novel gustatory stimulus used in CTA experiments should elicit no or only mild neophobia. The CS intake is thus only slightly lower than the water intake. At a definite interval after CS presentation (up to several hours) the animal receives the US (in the most common case an i.p. injection of isotonic lithium chloride, 2% body weight), which elicits during 5 to 10 min symptoms of poisoning. The animals become flaccid, are stretched out on their bellies, show increased peristalsis, diarrhoea, and reduced reactivity to external stimuli. The symptoms culminate after 1–2 h and disappear after 5–10 h. Water consumption almost returns to control level on the next day. Since more severe poisoning can elicit longer lasting disturbance, water consumption should be followed for several days till full normalization. New presentation of the CS taste reveals a marked decrease of fluid consumption against the first presentation.

Consequences of consumption of a novel flavour can thus be manifested during its second presentation (retention test) by a continuum of responses ranging from increased (AN) to decreased (CTA) intake of the CS fluid. Statistical evaluation can compare the fluid intake during the first and second presentation of the CS. A more reliable comparison can be made between the second presentation of the CS in poisoned (experimental) and non-poisoned (control) animals. This comparison takes into account the possible increase of fluid intake due to attenuation of neophobia (Fig. 2.1).

The main limitation of the single bottle procedures is the inherent conflict between motivation and aversion which determines the absolute volume of fluid consumption. With strong CTAs fluid intake may drop to zero and differences between their intensities may be obscured by the floor effect. In this case it is necessary to distinguish between CTAs of various intensities by examining their resistance to extinction.

2.1.2 Multiple bottle procedures

Some of the above difficulties are removed in the two-bottle or multiple bottle techniques which give the animal the opportunity to choose between the familiar fluid (water) and the novel fluid, or between water and several other fluids. The animal is not forced to drink the novel fluid and neophobia (expressed as the ratio of the volume of the novel fluid to total fluid intake) is usually expressed more. In CTA experiments the two-bottle procedures are usually employed only during retention testing, while both bottles contain water or the CS fluid during CTA acquisition. CTA is expressed by the saccharin intake/total fluid intake ratio. A non-poisoned group must be used as control and statistical comparisons should be made between the CS/(CS + water) ratios in the experimental and control groups. The two-bottle method can reveal weaker CTAs than the single bottle technique (Dragoin *et al.* 1971; Grote and Brown 1971) but fails to discriminate between strong aversions. As pointed out by Batsell and Best (1993) single bottle procedures are better suited for the latter task.

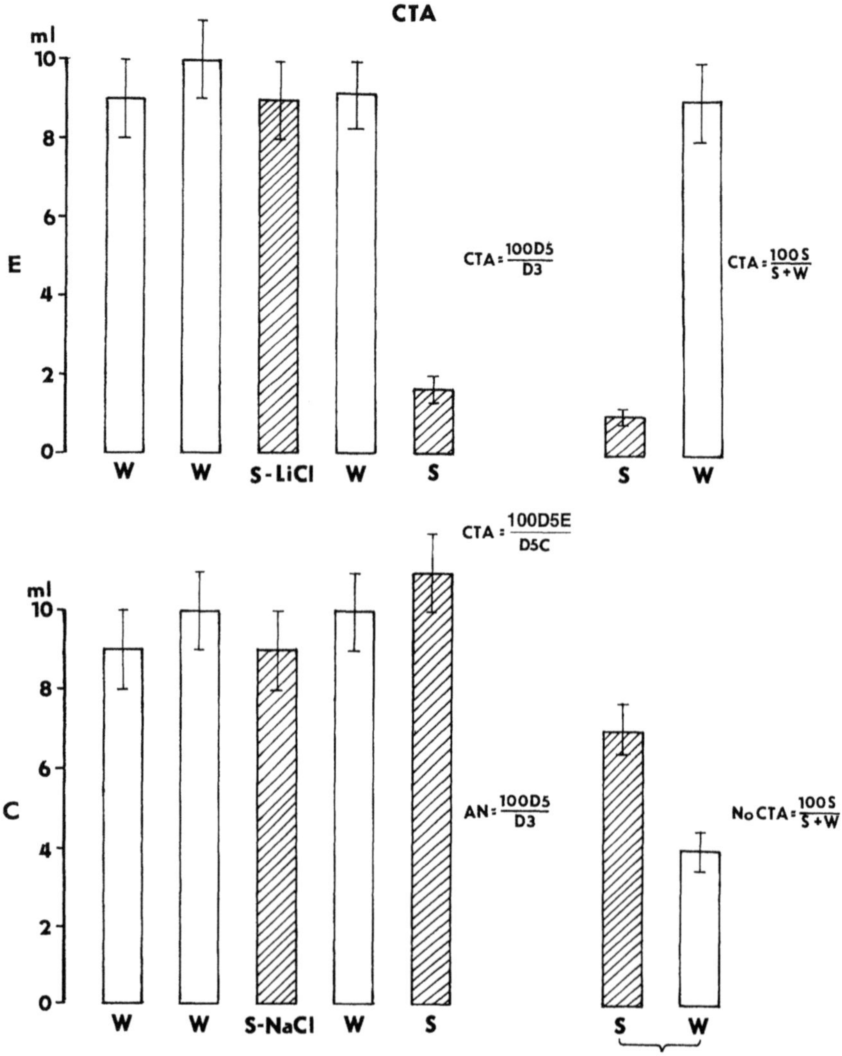

Fig. 2.1 The CTA paradigm. Example of illustrative results in control (C, $n=10$) and experimental (E, $n=10$) groups. Left: single choice procedure. Abscissa: days of the experiment. The rats were allowed 15 min access to water (W, empty columns) on Days 1, 2, and 4 or to 0.1 % sodium saccharin (S, shaded columns) on Days 3 and 5. Saccharin drinking on Day 3 was followed after 15 min by intraperitoneal injection of 2% body weight of 0.15 M NaCl in group C or of 0.15 M LiCl in group E. Ordinate: mean ($\pm$SEM) consumption of the respective fluids. Saccharin intake on Day 5 expressed in percentages of saccharin intake on Day 3 indicates AN in C and CTA in E. A better CTA estimate is obtained when expressing saccharin intake on Day 5 in E in percentages of saccharin intake on day 5 in C. Right: two-choice procedure. On Day 5 the rats are offered two or more bottles, half of them containing water and half saccharin. Percentage of saccharin from the total fluid intake is calculated for each rat and the mean ($\pm$SEM) value is used as an index of CTA in group E and compared with the corresponding value in group C.

An undesirable feature of the two-bottle technique is the chance factor affecting the rat's access to the two spouts. This is a particularly important source of variability when short access to the drinking bottle is used. The animal may by chance approach the water bottle and drink from it continuously without sampling the other fluid. This can be avoided when the fluids are offered not in two large pipettes but in a number of small (e.g. 2 ml) equally accessible pipettes filled alternately with the two liquids (Bures *et al.* 1988). A rat with a mean fluid intake of 10 ml must visit at least five such pipettes to satiate its thirst. This provides a more equal exposure of the animal to both fluids the consumption of which is obtained by the cumulative sums at all visited pipettes containing the same fluid.

Another modification of the two-bottle technique requires the animal to consume the fluids sequentially, e.g. by offering them only one spout at a time. This can be done by using the retractable spouts which can be alternately presented at two adjacent openings in the front wall of the drinking box (Mamedov *et al.* 1987). The position bias can be completely eliminated when the two fluids are alternately presented at the same wall opening. This can be achieved either by a mechanical switching device which rotates two spouts so that only one is always presented at the same site (Brozek *et al.* 1979*a*) or by perfusing a small space (10 μl) at the tip of the drinking spout with one of the two fluids (Brozek *et al.* 1979*b*). In both cases, drinking must be measured by the number of licks emitted during exposure to one or the other flavour (see Section 2.1.4 and Fig. 2.2 for details).

2.1.3 Taste reactivity testing

The single bottle and multiple bottle procedures measure spontaneous intake of fluids which does not necessarily reflect the palatability changes induced by CTA training. The taste reactivity test introduced by Grill and Norgren (1978*a*) and further developed in a number of follow-up studies (Parker 1980; Grill and Berridge 1985; Spector *et al.* 1988; Ossenkopp and Eckel 1995) makes it possible to address the palatability issue. The animal is implanted with a flexible intraoral cannula (Parker 1980) connected through a piece of injection needle tubing fixed to the skull bones to an infusion pump. Intraoral infusions lasting up to 3 min are applied at the rate of 10 μl/s. Orofacial and somatic responses to the taste stimulus application are videotaped through a mirror placed at a 45° angle beneath the glass floor of the observation chamber.

Characteristic responses observed during infusion of palatable (sucrose, saccharin) and unpalatable (quinine) fluids are summarized in Table 2.1. The ingestive sequence includes rhythmic movements of the mouth, tongue protrusions, and lateral tongue protrusions. The aversive sequence consists of gapes, chin rubs against the floor, head shakes with fluid expulsion, paw wipes, and paw flails (to remove fluid from the fur). The above aversive responses are supplemented by the passive drip manifested by accumulation of water around the mouth followed by dripping to the floor. Infusion of sucrose into the oral cavity of naive rats elicits ingestive responses but aversive responses in rats with CTA to sucrose after sucrose–LiCl pairing (Berridge *et al.* 1981). On the other hand, CTA to sucrose paired with ingestion of large quantities of lactose (Pelchat *et al.* 1983) elicits a clear decrease of spontaneous sucrose intake but does not change the ingestive response to intraoral infusion of sucrose. Similar dissociation of fluid intake and taste reactivity was reported by Parker (1982) in CTAs elicited by sucrose–LiCl and sucrose–amphetamine pairings. Under conditions of equally strong suppression of sucrose intake, only the LiCl-induced CTA was accompanied by the palatability shift.

Table 2.1 Characteristic responses evaluated in the taste reactivity testing

Ingestive responses, mediating continuous fluid intake

Rhythmical mouth movements
—symmetrical lowering of the mandible at a 6 Hz rate

Tongue protrusions
—midline extensions of the tongue over the upper lip

Lateral tongue protrusions
—lateral extensions of the tongue combined with unilateral retraction of the upper lip

Paw licks
—midline extension of the tongue directed to the forepaws

Aversive responses, mediating ejection of the infused fluid

Gapes
—lowering of the mandible leading to exposure of lower incisors, retraction of the corners of the mouth, and fluid expulsion

Chin rubs
—bringing the chin in contact with the floor or walls and depositing the fluid on these surfaces

Head shakes
—rapid side-to-side movements of the head with open mouth leading to removal of the fluid from the mouth

Paw wipes
—a rearing rat uses forepaws to remove fluid from the mouth

Forelimb flails
—rapid horizontal movements of the forelimbs removing fluid from the fur

Passive drip
—in absence of ingestive responses fluid accumulating in the mouth drips on the floor

While taste reactivity tests are usually employed for estimating the intensity of well-established CTAs, they can also be used for dynamic monitoring of CTA acquisition. This technique was employed by Spector *et al.* (1988) and Ossenkopp and Eckel (1995) for rapid formation of CTA. The rat with implanted intraoral cannula received i.p. injection of LiCl followed immediately by a series of brief (30 s) infusions of a novel taste (e.g. sucrose) repeated at 5 min intervals. While the first infusions elicited almost exclusively ingestive responses, their frequency started to decrease during the third infusion (10 min after LiCl injection). At the same time incidence of aversive responses started to increase. After CS infusion 25 min after LiCl administration the rat's response to intraorally applied taste changed from an ingestive pattern to an aversive one. The fact that the response to sucrose remained aversive even during CS infusions applied after dissipation of the LiCl effect several hours or days later demonstrates that the reactivity change was not due to poisoning enhanced neophobia, but to formation and retention of CTA. The associative nature of the observed reactivity change was borne out by the fact that they only appeared after repeated pairing of taste with the LiCl malaise. When the first sucrose infusion was applied 20 min after LiCl injection, it elicited an ingestive response which changed to an aversive one only after several further pairings of the sucrose infusions with the visceral symptoms of poisoning. An important advantage of the method is that it makes it possible to establish the time when a neutral or positive CS acquires the properties of an aversive stimulus. Exact determination of the corresponding time is important for electrophysiological and neurochemical studies purporting to detect the neural correlates of the associative change.

2.1.4 Licking behaviour

Closely related to taste reactivity testing is the analysis of the animal's response to the contact with the drinking spout containing palatable, neutral, or aversive fluids. In the standard laboratory environment the rat's drinking at the water spout (Weijnen and Mendelson 1977) can be recorded by the so-called drinkometers which monitor the contact of the tongue with the spout either by passing a low electric current (about 1 μA) between the animal standing on a metal floor and the spout or by photoelectric detection of the movement of the tongue across the gap between the hole in the cage wall and the orifice of the external spout. Licking is recorded in the form of rectangular pulses, the duration of which corresponds to tongue protrusion beyond the level of the sensor. The record has a highly regular frequency of 5–6 Hz with lick duration of about 60 ms. With each lick about 5 μl of the fluid are transported from the spout into the mouth. By continuously licking, the animal may drink 1.5–1.8 ml/min. When the water in the spout is suddenly replaced by the aversive fluid a rat with a strong CTA needs a single lick to recognize the changed condition and to cease licking (Halpern and Tapper 1971). This means that activation of the taste buds, conduction of the nerve volley through the gustatory pathway, analysis of the afferent signal, its comparison with the information stored in the memory, and the decision to continue or discontinue licking must be completed within a single lick cycle, i.e. during 170 ms. The actual time limit is still shorter because each lick starts about 50 ms before it is photoelectrically detected and because activation of the gustatory receptors requires about 30 ms. This analysis suggests that the memory readout and elaboration of the command to stop licking takes place between 50 and 100 ms after the first contact of the tongue with the aversive fluid. In this time bracket it is necessary to search the electrophysiological correlates of the underlying neural processes.

The number of licks required for taste recognition can be tested at the onset of a bout of licks, e.g. when the animal approaches a newly presented spout or during a sudden change of the taste of the consumed liquid. The two situations are different, because in the first case the gustatory stimulus triggers an equiprobable appetitive or aversive response whereas in the second it either approves continuation of consummation or leads to its cessation. Since taste belongs to spatially non-oriented sensory organs it cannot be used for simultaneous discrimination studies. This property is further enhanced by taste presentation in discrete licks. The closest approximation to a simultaneous gustatory discrimination is the procedure developed by Brozek, Buresova and Bures (1979*a*) which presents to the animal simultaneously two different fluids delivered through two drinking spouts placed 20 mm apart (Fig. 2.2). Solid state programming circuits make it possible to switch the spouts within a 50 ms interval. To prevent extinction of the aversive trace the gustatory stimulus is mildly toxic, e.g. an isotonic solution of LiCl. In the typical experiment a rat with established CTA to the salty taste of LiCl licks at the water spout. When after 30 to 40 licks spouts are switched, the first lick at the aversive fluid appears after a train of water licks which should be stopped as soon as the aversive taste is recognized. After one or two LiCl licks the animal moves to the spout now containing water and continues licking there. The critical events in this gustatory discrimination procedure are (a) the last water lick, (b) the first LiCl lick, (c) the last LiCl lick, (d) the first water lick at the other spout. Comparison of the electrical correlates of (a) and (b) in naive rats (which lick water and LiCl indiscriminately) reveals differences between the sensory properties of the stimuli. Comparison of the responses to (b) in naive and CTA trained rats should reflect the memory retrieval process. The response to (c) is followed by the command to leave the currently used spout. Finally, comparison of (a) and (d) reflects the difference between a habituated and a dishabituated response to the same signal.

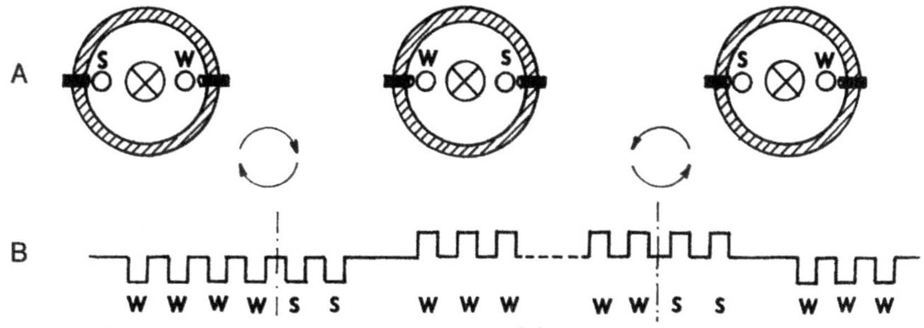

Fig. 2.2 The gustatory discrimination apparatus. (A) Front view. Two round openings in the front wall of the experimental box provide access to two drinking spouts connected to infusion pumps delivering water and salt, respectively. The spouts mounted in a plastic cylinder can be rapidly (50 ms) switched between the two wall openings by a reversive electromotor. Licking is recorded by a photoelectric system consisting of a light source and phototransistors monitoring the movement of the tongue across the gap between the wall and the spout. (B) Recording of spout switching and of the water and salt licks. Licking in the left and right openings is indicated by upward and downward oriented deflections. Record starts with licking water (W) from the right spout. After the spouts are switched (interrupted vertical line), the rat continuing licking at the right spout is now exposed to salt (S). After two S licks the rat moves to the left spout now delivering water and stays there until the spouts are switched again.

The apparatus and its various modifications were used in electrophysiological experiments examining neuronal activity in CTA related brain centres for lick-related modulation (Buresova *et al.* 1979) or attempting to disrupt CTA retrieval by lick synchronized stimulation of relevant centres (Brozek *et al.* 1979*b*).

2.1.5 CTA acquired without ingestion

In a typical CTA experiment taste is presented during spontaneous consummation of flavoured food or fluid. Thus, although CTA is considered an example of classical (Pavlovian) conditioning, exposure to the gustatory CS occurs as part of an instrumental or operant behaviour leading to ingestion of the stimulus. The spontaneity of ingestion is partly eliminated when the taste is applied by intraoral infusion, but even in this case the animal can avoid ingestion by emitting active or passive aversive responses. In particular, the passive drip shows that the infused fluid was not swallowed. Gustatory stimuli applied by intraoral infusion are less efficient than those presented during spontaneous drinking (Domjan and Wilson 1972; Buresova and Bures 1973), but can nevertheless produce CTA in intact animals. In fact, CTA can be acquired in the absence of any consummatory responses, e.g. when swallowing was suppressed by curarization (Domjan and Wilson 1972). In this case, CS is only acting on the taste receptors of the oral cavity and does not enter the intestinal tract. This eliminates possible stimulation of visceral receptors by the CS and activation of oral receptors by CS regurgitation of the ingested flavour or by penetration of the flavour into the circulation.

The latter possibility can be deliberately exploited when inducing CTA by systemic administration of high concentrations of the taste substance. The taste buds can be stimulated not only by flavours present in the oral cavity but also by those transported through the blood. This so-called 'vascular taste' was used already in the 1930s (Fishberg *et al.* 1933) for measurement of the circulation time from cubital vein through right ventricle, lungs, left ventricle, and aorta to carotid artery. CTA to vascular taste was demonstrated for i.v. applied saccharin by Bradley and Mistretta (1971) and for i.p. applied saccharin by Baum *et al.* (1974), Buresova and Bures (1977) and Bellingham and Lloyd (1987). The latter authors showed that CTA strength increases with the CS–US delay and attains a maximum at 2 h. These results confirmed by Bures and Buresova (1989) suggest that the taste elicited by i.p. injected saccharin is detected only after a delay of about 30 min. Low efficiency of CTA acquisition with short CS–US intervals may be due to the fact that CS perception does not precede but follows the US. Such reverse conditioning is further confounded by the so-called 'medicine effect', i.e. by positive labelling of the taste stimuli coinciding with the recovery from poisoning.

The fact that CTA can be induced by systemically applied flavours raises the possibility that some aversion inducing substances used as US can reach blood concentrations eliciting gustatory sensations which may serve at the same time as CS. This seems to be true in the case of alcohol the parenteral application of which induces CTA manifested by reduced alcohol intake (Cunningham 1978; Miceli *et al.* 1980).

2.2 CTA-related phenomena

2.2.1 Conditioned taste preferences

Whereas avoidance of highly toxic food is vitally important, food preferences are usually formed under less dramatic circumstances. The best known examples are preferences for food which counteracts a dietary deficiency, e.g. for thiamine containing diet in thiamine deficient rats (Rodgers 1967; Capretta 1977) or for salt in rats maintained on sodium deficient diet (Sakai *et al.* 1987; Schulkin 1991). A more general paradigm compares consumption of two flavours one of which (CS+) is combined with sucrose whereas the other (CS-) is unaltered (Reilly *et al.* 1993). Preference for the CS+ flavour can be demonstrated in a two-bottle test after sucrose is omitted. An analogous procedure establishes taste preference by associating one taste (CS-) with poisoning and the other taste (CS+) with saline. Simultaneous presentation of CS+ and CS- in a two-bottle extinction test shows preference for CS+.

CS+ and CS- can also be offered in the course of the same session (García *et al.* 1974), CS- before application of the toxin (apomorphine) and CS+ shortly before recuperation from illness. During five presentations of the above sequence of stimuli CS- intake gradually decreased and CS+ intake increased. Subsequent two-bottle testing showed clear preference for the CS+.

It must be pointed out that a taste preference can also be described as aversion for the deficient diet. To prove that a specific preference was formed for CS+ requires that it is preferred not only against the CS- (i.e. the deficient diet), but also against other novel flavours or non-deficient diets.

2.2.2 Changing food classification

CTA and conditioned taste preferences are extremes of a continuum proposed by Rozin and Kalat (1971). According to these authors foods can be classified as familiar-aversive (least preferred), novel and familiar-safe (most preferred). The latter category is not preferred more than tastes which have been paired with the recovery from illness (Zahorik 1977).

The above classification is continuously changing. In fact, a novel food is novel only during its first presentation and is then classified as familiar-aversive or familiar-safe. More stable are the latter categories, but they should also change according to circumstances. Thus a taste which is no longer signalling poison should be reclassified as familiar-safe and a taste signalling safety should become familiar-aversive when associated with poisoning. Some processes which participate in the above modifications are listed below.

2.2.3 CTA extinction

When a gustatory CS eliciting a conditioned response (i.e. an aversive oromotor response and reduced consumption of the flavoured food or fluids) is presented without subsequent exposure to the US (i.e. poisoning), it gradually ceases to elicit the aversive response. Extinction is usually elicited using a single bottle procedure allowing presentation of the aversive CS as the only source of liquid. As shown by Grote and Brown (1973) extinction proceeds faster in severely water deprived rats. Slow extinction takes place even with the two-bottle procedure when the aversive fluid is present simultaneously with a familiar safe taste (usually water). Although under such conditions the animals do not need to consume the averted taste, sampling small amounts of the presumably toxic food is a part of long-term feeding strategy which continuously tests persistence of the aversive properties of all potential sources of food. Extinction can be elicited even when the aversive taste is presented systemically, e.g. by i.p. injection (Baum *et al.* 1974). Extinction is manifested by return of the CS flavour preference to the level corresponding to the familiar-safe category and by disappearance of the aversive response in the taste reactivity tests.

2.2.4 Attenuation of neophobia

Neophobia is a cautious attitude to novel stimuli, manifested by faster cessation of drinking and withdrawal from the unknown gustatory taste. Repeated presentation of the gustatory CS gradually leads to attenuation of this neophobic response by a process akin to extinction. Similar to extinction, AN leads to classification of the respective taste as familiar-safe. AN is proportional to the amount of the consumed CS and depends, therefore, on the conditions of flavour administration (one-bottle or two-bottle) and on the motivation of the animal (Domjan 1977).

2.2.5 Sensitization

Poisoning occurring in the context of drinking increases rejection of novel flavours presented in a similar situation (Rozin 1968; Domjan 1977). This enhanced neophobia applies to responses to novel tastes but does not affect responses to familiar-aversive or familiar-safe flavours. It is particularly expressed for edibles resembling the food eaten shortly before the poisoning. Poisoning occurring in the absence of prior consumption of edibles elicits sensitization lasting only for the duration of the intoxication symptoms, but the enhanced neophobia is evoked by a broad spectrum of taste stimuli.

2.2.6 Conditioned inhibition and higher-order conditioning

A new taste applied shortly after presentation of the aversive gustatory CS but not followed by an aversive visceral state will assume inhibitory significance, i.e. will be later preferred in the two-bottle choice situation (Best 1975). Timing of the exposures to the novel gustatory CS and to the aversive CS is critical in these experiments. Rats receiving intraoral infusion of the new taste followed by similar infusion of the aversive taste acquire an aversion to the new taste by the mechanism of second-order conditioning (Bond and DiGiusto 1976).

2.2.7 Latent inhibition

A taste stimulus which has been repeatedly presented to the animal in absence of any aversive consequences requires more pairings with poisoning to elicit CTA of comparable intensity (Domjan 1972; Elkins 1973; Albert and Ayres 1989; De La Casa and Lubow 1995). Such decrease of conditioning performance called latent inhibition is explained by interference with the processing of stimuli which, in the past, proved to be insignificant. According to this view, latent inhibition results in a failure to acquire new association to such stimuli. An alternative view asserts that latent inhibition is due to a retrieval failure, i.e. that association of the pre-exposed CS to the US proceeds normally but that during retrieval the earlier CS–no consequence association competes with the later CS–poisoning association (Bakner *et al.* 1991; Kraemer *et al.* 1991; Batsell and Best 1992).

2.2.8 Overshadowing

When two cues of markedly different salience are presented as a compound CS, the less salient stimulus forms a much weaker CTA than it would in the absence of the more salient cue, whereas CTA to the more salient stimulus remains undiminished. The stronger stimulus may overshadow the weaker one so markedly that no signs of conditioning of the latter stimulus can be revealed even by the most sensitive methods (Lindsay and Best 1973; Best and Meachum 1986).

2.2.9 Blocking

According to Kamin (1969) a well-established CS presented on later conditioning trials in compound with a new stimulus prevents or blocks the new stimulus from associating with the US. It is assumed that reliable predictability of the US by the CS reduces US associability with other cues which may serve as potential CS. Blocking was demonstrated in the CTA paradigm by Revusky (1971) and by Gustavson *et al.* (1992).

3 Neuroanatomy of CTA: lesion studies

For more than twenty years researchers in several laboratories have investigated the neural pathways involved in CTA. As we will review in this section, based mainly on results obtained in laboratory rodents, taste fibres carry information about the flavour of a meal to the nucleus tractus solitarius (NTS) in the brainstem. Visceral fibres carrying the feedback from food ingestion converge to the same brainstem and midbrain areas. When a meal is toxic, it may cause gastrointestinal irritation that is carried by the vagus nerve to the NTS; alternatively, if the toxin is absorbed into the bloodstream, it is carried to the area postrema (AP) in the same region of the brainstem. The same mechanisms control both the vomiting reflexes that eject the toxic food from the stomach, and the CTA that blocks ingestion of the toxic food on the next encounter. For John García these anatomical relationships provide a convenient feedback loop for the processing of information regarding the gustatory qualities of food, and the subsequent visceral effects following ingestion of a poisoned meal (García *et al.* 1974, 1985; Coil *et al.* 1978).

This chapter describes the effects of irreversible brain lesions on the acquisition and retention of CTA and reviews functional significance of various brain regions for CTA formation. The brain sites relevant to CTA mechanisms are shown in Fig. 3.1.

3.1 Area postrema

AP is a small circumventricular organ lacking the blood–brain barrier and located on the midline floor of the caudal fourth ventricle. Neuroanatomical studies have shown that the AP receives peripheral inputs via vagal afferent fibres (Contreras, *et al* 1982; Kalia and Sullivan 1982) and central inputs mainly from neurons in the paraventricular and dorsomedial hypothalamic nuclei (Shapiro and Miselis 1985). The AP projects to the NTS, PBN, dorsal motor nucleus of the vagus, nucleus ambiguus, and other brainstem structures (van der Kooy and Koda 1983; Shapiro and Miselis 1985).

Thus, the AP has been implicated in neural as well as in humoral integration of visceral sensation, and considered as an emetic chemoreceptor trigger zone (Borison and Wang 1953). In the CTA paradigm, AP presumably detects the US either directly as a toxic substance present in the blood or indirectly via visceral afferent nerves.

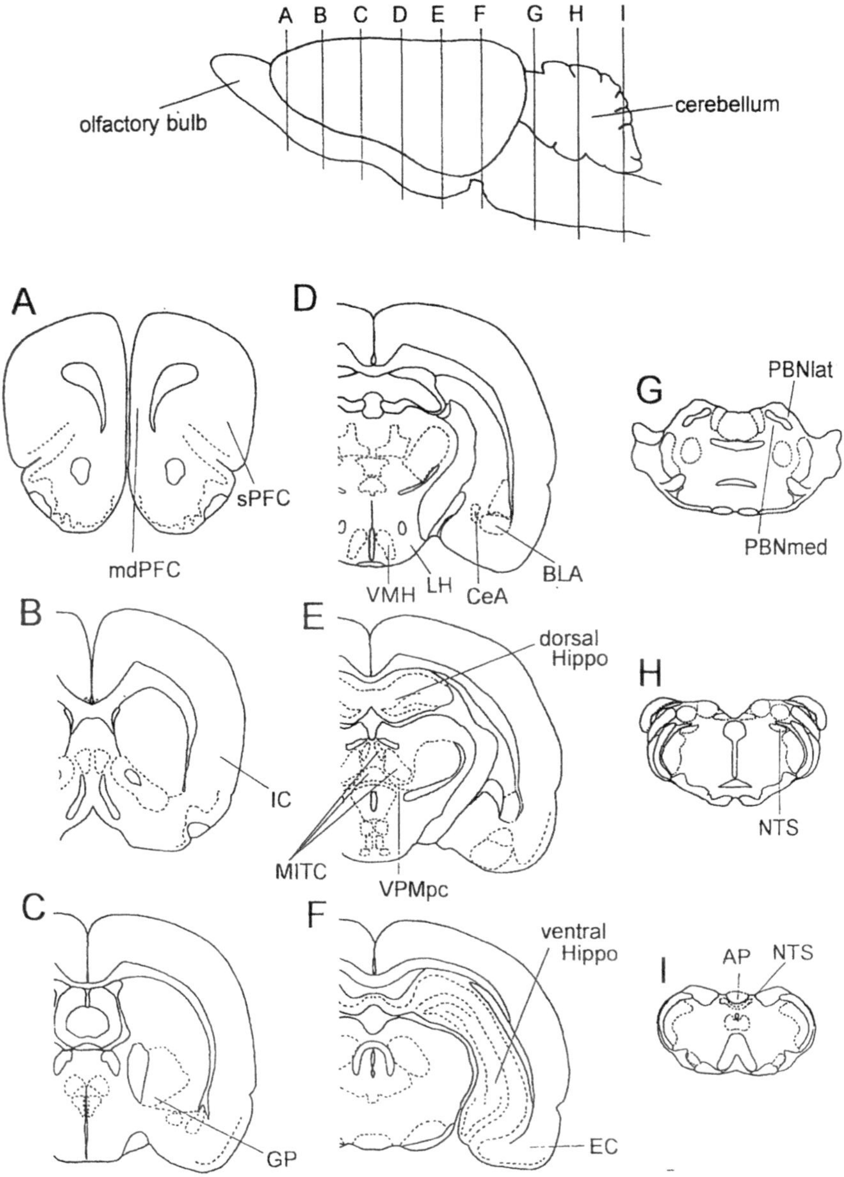

Fig. 3.1 Brain sites related to gustatory and general visceral senses and feeding behaviour in the rat. Abbreviations: AP, area postrema; BLA, basolateral nucleus of the amygdala; CeA, central nucleus of the amygdala; EC, entorhinal cortex; GP, globus pallidus; Hippo, hippocampus; IC, insular cortex; LH, lateral hypothalamic area; MeA, medial nucleus of the amygdala; MITC, midline and intralaminar thalamic complex; NTS, nucleus of tractus solitarius; sPFC, sulcal prefrontal cortex; mdPFC, mediodorsal prefrontal cortex; PBN, parabrachial nucleus; PBNlat, lateral part of PBN; PBNmed, medial part of PBN; VMH, ventromedial hypothalamic nucleus; VPMpc, parvocellular part of ventral posterior medial nucleus of thalamus.

Many lesion experiments have been performed to examine the role of the AP in the action of various USs. When the US is LiCl (Ritter *et al* 1980; Rabin *et al* 1983*a*; Ladowsky and Ossenkopp 1986; Kosten and Contreras 1989; Curtis *et al* 1994), scopolamine (Berger *et al* 1973; Ossenkopp *et al* 1986; Kosten and Contreras 1989), gamma radiation (Ossenkopp 1983a; Rabin *et al* 1983*a*; Ossenkopp and Giugno 1985), or histamine (Rabin *et al* 1983*a*), AP-lesioned rats exhibit striking deficits in the ability to develop CTA. Such rats, however, develop CTA to other USs such as amphetamine (Berger *et al* 1973; Rabin *et al* 1987), nicotine (Ossenkopp and Giugno 1990) and ethanol (Stewart *et al* 1988; Hunt *et al* 1987). Rabin *et al* (1983b) showed that the AP lesions made after the acquisition of CTA to sucrose followed by LiCl or gamma radiation had no effect on the retention of the sucrose aversion. This results indicate that AP is important only in the acquisition phase of CTA, but is not indispensable for long term persistence of CTA memories and their retrieval.

Participation of the AP in the mediation of the US effects may depend on the experimental procedures used. For example, AP lesions did not disrupt CTA elicited by intragastric administration of copper sulfate, but impaired CTA acquisition when the same toxin was administered intravenously (Coil and Norgren 1981). When motion sickness was used as the US, AP lesioned rats either did (Ossenkopp 1983*b*; Sutton *et al* 1988) or did not (Gallo *et al.* 1991) acquire CTA depending on the type of vestibular stimulation employed. Similarly, AP damage does not prevent scopolamine-induced CTA in some procedures (Gallo *et al.* 1990). Amphetamine-induced CTA may require participation of the AP at low concentrations since AP lesioned rats cannot establish CTA when the US is a combination of subthreshold irradiation and subthreshold amphetamine (Rabin *et al* 1987). Acquisition of CTA following body-only irradiation is totally mediated by the AP, while CTA following head-only irradiation is mediated partially by the AP (Rabin *et al* 1984). When using hypertonic NaCl as the US, Arnedo *et al.* (1990; 1991) and Arnedo *et al.* (1993) established that the AP is not required for 'short-term taste aversion learning', which is induced by non-delayed or concurrent presentation of the gustatory CS with the hypertonic NaCl. On the other hand, circulatory system and structures such as the AP play important roles in establishing 'long-term aversion learning', characterized by the conventional delay between presentations of the CS and US. They also suggested that the vagus nerve is critically involved in the short-term but not in the long-term learning. It is conceivable that CTAs formed with different protocols and based on different USs are mediated by different neurobiological mechanisms.

3.2 Nucleus tractus solitarius

The NTS (the first gustatory relay) receives heavy visceral input from different branches of the vagus nerve (which is sensitive to gastrointestinal irritating toxins), as well as input from AP (which is sensitive to blood-transported toxins), and from the vestibular system (which is sensitive to nausea-causing motion). These inputs terminate in the caudal two-thirds of the NTS. The NTS also receives primary taste afferents from the entire tongue via nerves VII and IX, from the soft palate via nerve VII and from the larynx and pharynx, via nerve X. These nerves terminate in a rostral to caudal sequence within the NTS. Gustatory neurons are located in the rostral third of the NTS (gustatory zone) which projects to the medial part of the pontine parabrachial nucleus (PBN), while the caudal NTS (general visceral zone) projects to the lateral part of the PBN.

According to Shimura *et al.* (1997a) and Grigson *et al.* (1997a,b), rats with electrolytic lesions of the rostral NTS exhibited consistent deficits in responding to sapid stimuli. For example, licking 1.0 M sucrose did not differ from licking water, and the addition of normally aversive substances reduced intake only about half as much as before lesion. When presentation of 0.3 M alanine was paired with injection of LiCl, lesioned rats learned to avoid alanine. These results suggest that rats with NTS lesions show a marked impairment of gustatory preference and aversion, but can still use taste cues for learned aversions. It is conceivable that the residual taste neurons in the caudal NTS are sufficient for gross recognition of taste or no taste, while the removed taste neurons in the rostral NTS are important for making fine discriminations between different intensities of various flavours. After lesions in the PBN, on the contrary, rats exhibit only moderate changes in gustatory preference, but cannot learn taste aversion. This suggests a more important role of the PBN in the associative rather than in the sensory aspects of gustatory functions.

3.3 Pontine parabrachial nucleus

The PBN is the second-order relay nucleus of the taste pathway in the rat (Norgren and Leonard 1971). Axons of neurons in the gustatory zone in the rostral part of the NTS ascend ipsilaterally through the reticular formation to the PBN in the caudal pons (Norgren 1984). The PBN is also a principal target of neurons in the caudal general visceral zone of the NTS (Cechetto 1987). Anatomical studies (Norgren 1978; Herbert *et al.* 1990) suggest that these two parts of the NTS project to different subnuclei in the PBN, i.e. the majority of neurons from the gustatory zone of the NTS project to the medial subnuclei of the PBN and from the visceral zone of the NTS to the lateral subnuclei of the PBN. Hermann *et al.* (1983) have shown that first-order visceral and gustatory afferents project to separate regions of the NTS and do not converge upon the same neurons within this structure, but that both gustatory and hepatic divisions of the NTS send efferents to a confined region of the PBN at posterior levels besides the aforementioned separate recipient sites, suggesting convergence of abdominal–vagal and gustatory afferents. The PBN also receives projections from the area postrema (van der Kooy and Koda 1983; Shapiro and Miselis 1985) which is known to be activated by various emetic stimuli including LiCl (Tsukamoto and Adachi 1994).

There are two main projections from the PBN. The first one ends in ventral forebrain structures, including the amygdala, lateral hypothalamus, substantia innominata, and bed nucleus of the stria terminalis. The second projection ascends in the central tegmental bundle to the posterior thalamus, zones identified as relay sites for gustatory, somatosensory lingual, and general visceral afferents (Norgren 1984).

Lesions of the PBN disrupt acquisition and retention of CTA more effectively than lesions of other brain sites including the thalamic gustatory area, cortical gustatory area, entorhinal cortex, amygdala, hippocampus, lateral hypothalamic area, and ventromedial hypothalamic nucleus (Fig. 3.2) (Yamamoto *et al.* 1995*a*). Several investigators have reported that bilateral ibotenic acid lesions (Scalera *et al.* 1995; Yamamoto *et al.* 1991, 1995*a*) of the whole PBN, which invade the taste and visceral recipient zones, severely impair the ability of rats to acquire CTA. Electrolytic lesions confined to the medial part of PBN (PBNmed), that receives taste information about the gustatory CS, disrupt acquisition of CTA (Di Lorenzo 1988*a*; Reilly *et al.* 1993*a*). Flynn *et al.* (1991) and Spector *et al.* (1992) demonstrated this disruption by analysing consummatory oromotor responses (taste reactivity test) to an intraorally infused CS that had been paired with LiCl injection. More lateral lesions which included the lateral part of PBN (PBNlat) impaired CTA induced by intragastric administration of LiCl (Aguero *et al.* 1993*a*) or electrical stimulation of the area postrema (Aguero *et al.* 1993*b*). Sakai *et al.* (1994) destroyed the PBNmed and PBNlat separately to examine the role of these areas on acquisition and retention of CTA to saccharin, and found that the rats with lesions of the PBNmed did not show neophobia to saccharin (CS) and failed to establish CTA to the CS paired with LiCl. This might have been caused by blockade of transmission of taste quality information. Rats with lesions of the PBNlat showed neophobia, but failed to establish CTA. Further, Sakai *et al.* (1994) found that retention of CTA is impaired severely by the PBNmed lesions but only slightly by PBNlat lesions. These results suggest that taste quality information of saccharin transmitted through the PBNmed is important for both acquisition and retrieval of CTA, but that the PBNlat, which may be concerned with processing of general

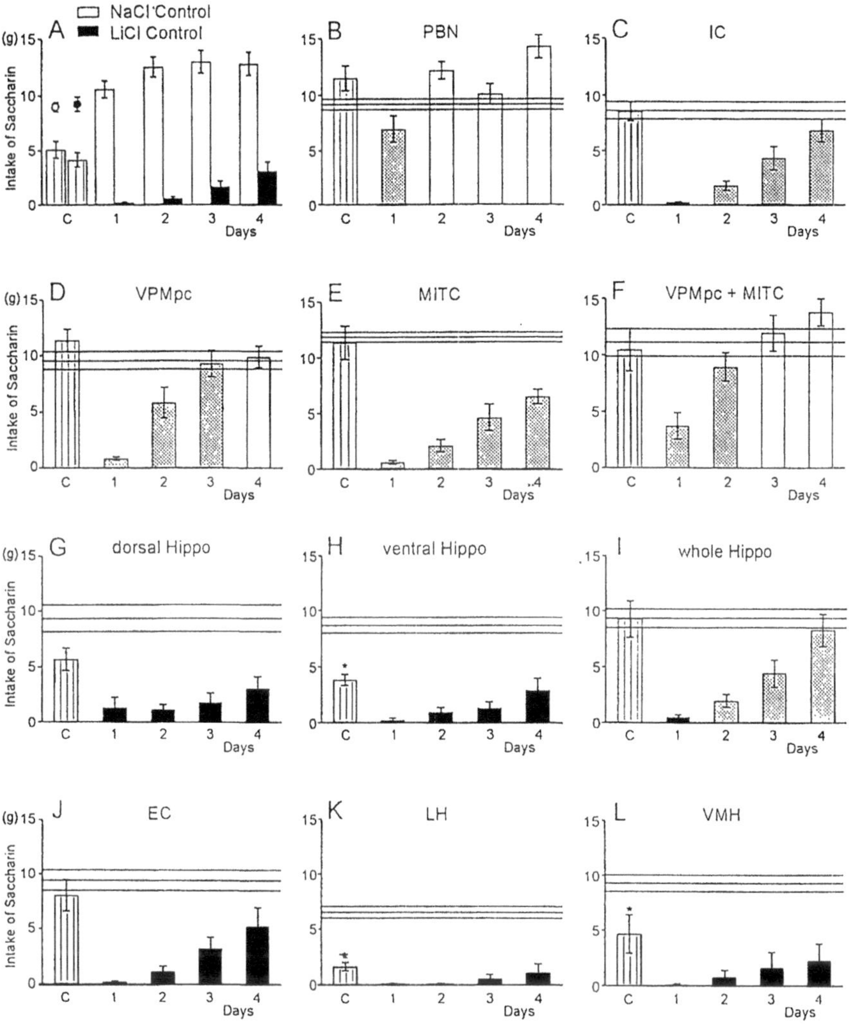

Fig. 3.2 Acquisition and retention of conditioned taste aversion to 0.2% sodium saccharin in control and experimental groups. The groups consist of NaCl-control and LiCl-control rats, and rats with lesions of PBN, IC, VPMpc, MITC, both VPMpc and MITC (VPMpc + MITC), dorsal hippocampus (dorsal Hippo), ventral hippocampus (ventral Hippo), whole hippocampus (whole Hippo), EC, LH, and VMH. Each column shows the mean ±SEM intake of 0.2% saccharin solution for 20 min on conditioning day (C, indicated by striped column) and subsequent 4 days (indicated by open column for NaCl-control rats and by solid column for LiCl-control rats). The open and solid squares with vertical lines in graph A, and the horizontal lines in other graphs, show the mean ±SEM of postoperative preconditioning water intake. The shaded column indicates that the mean volume of intake is significantly smaller than that of NaCl-control rats and larger than that of LiCl-control rats on the corresponding days. The solid column indicates that the intake is significantly smaller than that of NaCl-control rats, but is not different from that of LiCl-control rats. The asterisk indicates that saccharin intake on the conditioning day is significantly lower than the water intake. Modified from Yamamoto *et al.* (1995*a*).

visceral signals and taste–visceral association, is only crucial for acquisition of CTA (Aguero *et al.* 1993*a*).

It is generally agreed that the initial integration of gustatory and visceral signals takes place in the PBN, and that the behavioural expression of acquisition and retention of CTA require participation of forebrain structures including the gustatory insular cortex and amygdala. Bures and Buresova (1990) believe that the CTA engram is formed in the PBN after this structure has received descending gustatory information from the gustatory insular cortex. Reilly *et al.* (1993*a*) suggest that their PBN lesions do not impair processing of taste stimuli or of general visceral stimuli, but interfere only with taste–visceral integration. Sakai *et al.* (1994) and Yamamoto *et al.* (1994*a,b*) propose that PBNmed lesions impair taste quality transmission and PBNlat lesions impair the function of the hedonically negative site where taste–visceral integration of CTA occurs. Taken together, it is not surprising that lesions of the PBN cause almost complete impairment of CTA formation.

3.4 Thalamus

The medial part of the PBN (gustatory zone) in rats projects to the parvocellular part of the ventral posterior medial nucleus of the thalamus (VPMpc) (Norgren 1984), and the more lateral portion of the PBN (general visceral zone) projects to the parvocellular part of the ventral posterior lateral nucleus of the thalamus (VPLpc) (Cechetto and Saper 1987). These thalamic contents in turn project to the insular cortex.

Earlier thalamic lesion studies in the rat employed different CTA paradigms, which makes it difficult to compare the results with each other. Loullis *et al.* (1978) presented rats with 0.125% saccharin followed by i.p. injection of LiCl. They found a loss of neophobia and marked attenuation of CTA formation in animals with electrolytic lesions of VPMpc. Lasiter (1985) also showed that an electrolytic lesion of the VPMpc markedly attenuated CTA learning; whereas normal rats acquired CTA to LiCl following the first conditioning trial, rats lacking VPMpc required approximately five conditioning trials to acquire similar LiCl aversions. Furthermore, he found that animals lacking both the olfactory bulbs and VPMpc could not acquire CTA to ingested LiCl, suggesting that olfactory cues associated with drinking play a significant role in CTA acquisition when the gustatory thalamic nuclei are destroyed. On the other hand, Flynn *et al.* (1991) reported that rats with VPMpc lesions could acquire CTA to alanine which had been paired with LiCl injections. They assessed CTA by the number of ingestive and aversive oromotor responses elicited by intraoral infusion of 1 ml of alanine rather than by using an intake test. However, they did not exclude the possibility that response could be mediated by non-gustatory (e.g. olfactory) cues.

Yamamoto (1994*b*) and Yamamoto *et al.* (1995*a*) showed that excitotoxic lesions of the VPMpc had little effect on the acquisition but markedly attenuated retention (or accelerated extinction) of CTAs (Fig. 3.2D), and that lesions of the midline and intralaminar thalamic complex (MITC) caused weak interference with acquisition and retention of CTAs (Fig. 3.2E). They also showed that lesions of both VPMpc and MITC caused more severe disruption of CTAs than isolated lesions of either structures (Fig. 3.2F). Neuroanatomical findings (Groenewegen and Berendse 1994) indicating that MITC receives projections from PBNlat (Saper and Loewy 1980; Groenewegen 1988) and sends axons to the basolateral nucleus of the amygdala (Ottersen and Ben-Ari 1979; Groenewegen *et al.* 1990) suggest that general visceral and hedonically negative information is sent to the amygdaloid nuclei other than the central nucleus via MITC.

Recently, Reilly and Pritchard (1996) reported that electrophysiologically guided small lesions confined to the VPMpc had no disruptive effects on the acquisition of CTAs. They claimed that the disruptive effects reported by previous workers were due to large lesions including not only VPMpc but adjacent areas such as the midline and parafascicular thalamic nuclei, the centromedial nuclei, and mediodorsal thalamic nuclei.

If their suggestion is correct, it may be plausible that the adjacent structures including the MITC or both MITC and VPMpc play more important roles in CTA formation than the VPMpc alone, and that a direct by-passing projection from the PBN to the insular cortex (IC) may take part of the function of the VPMpc. In accordance with this latter possibility, Yamamoto *et al.* (1995*a*) have shown that combined lesions of the thalamus and IC almost completely disrupted the acquisition of CTAs.

3.5 Hypothalamus

Yamamoto *et al.* (1995*a*) showed that ibotenic acid lesions of the hypothalamic nuclei regulating feeding behaviour, i.e. the lateral hypothalamic area (LH), known as the feeding centre, and the ventromedial hypothalamic nucleus (VMH), known as the satiety centre, had essentially no, or rather facilitating, effects on the formation of CTA (Fig. 3.2K,L). Weisman *et al.* (1972) also reported that rats with electrolytic lesions of the VMH displayed a stronger aversion to sweet milk paired with methylatropine and were more resistant to extinction than controls. Similarly, Kramer *et al.* (1983) demonstrated that hyperphagia-inducing knife cuts in the VMH of the rat did not alter aversive taste conditioning but that hypothalamic obesity did enhance CTAs. Gold and Proulx (1972) found that VMH-lesioned rats acquired an aversion to saccharin paired with apomorphine, but more slowly than controls did, and their aversion extinguished more rapidly.

On the other hand, Roth *et al.* (1973) showed that the LH-lesioned rats failed to learn to avoid a preferred flavour that signalled impending electric shock to the tongue, and that the LH rats retained a taste aversion acquired prior to lesioning but 70% of them failed to learn a new aversion. Although a further study is necessary, some of these discrepancies might be attributed to procedural differences: electrolytic lesions vs. excitotoxic lesions.

3.6 Globus pallidus

Globus pallidus (GP) is known to play important roles in the regulation of feeding. To examine the effect of pallidal microlesions on behavioural aspects of gustatory functions, Hernadi *et al.* (1997) made small, well-circumscribed lesions by microiontophoretic delivery of kainate into the ventromedial pallidal region of the rat, and found that these lesions prevented the animals from proper acquisition and retention of CTA to 0.01 M citric acid. They also found that lesioned animals displayed significantly stronger and persistent neophobia to the citric acid solution. De Luca and Monda (1986) showed similar results in rats with electrolytic lesions of the GP by using 0.1% sodium saccharin as the CS in the CTA paradigm. These results suggest that intrinsic neurons of the GP may play an important role in the attenuation of neophobia and in the acquisition, memory storage, and retrieval mechanisms of feeding-associated taste information.

3.7 Amygdala

The amygdala is a small collection of neurons forming a part of the limbic system located in the ventral temporal lobe. The mammalian amygdala can be divided into two main nuclear groups: a phylogenetically older group composed of the medial, central, and cortical nuclei, and a phylogenetically new group composed of the lateral and basal nuclei. Presumably, the basal nuclei originated by migration from the piriform cortex (Johnson 1923). Recently, a vast number of nuclei and subnuclei have been identified in the two original areas by electron microscopy. However, from a clinical and electrophysiological point of view, division of the amygdala into the above two areas is still considered appropriate (Richardson 1973). Both divisions have extensive interconnections with the hypothalamus: the corticomedial nucleus via the stria terminalis and the basolateral nuclei via the ventral amygdalofugal pathway. Inputs to the amygdala come from the orbital frontal cortex, prepiriform cortex, hypothalamus, thalamus, all sensory modalities, and from other parts of the limbic system (Richardson 1973). Furthermore, there is a direct projection from the central nucleus of the amygdala to the hypothalamus and brainstem areas that has been postulated to be mediating fear and anxiety (Davis 1994). In this regard, it has been reported that stimulation of the amygdaloid complex produces changes in the autonomic nervous system manifested by inhibition of respiration. Stimulation of the corticomedial nuclei elicits increase in blood pressure and heart rate, pupillary dilation and salivation. Projections from the central nucleus of the amygdala to lateral hypothalamus appear to be involved in the activation of the sympathetic autonomic nervous system during fear and anxiety (Krettek and Price 1978; Price and Amaud 1981; Le Doux *et al.* 1988). Both the amygdala and insular cortex project to a number of autonomic regulatory areas influencing autonomic activity, and are recipients of visceral afferent information (Norgren 1976; Kapp *et al.* 1985; Moga and Gray 1985).

The amygdala has been associated for some time with the motivational aspects of regulation of food and water intake. It has been demonstrated that amygdaloid stimulation produces changes in the reinforcing properties of food intake, i.e. stimulation of the dorsomedial amygdala increases food intake, stimulation of the basolateral or corticomedial amygdala produces inhibition of ongoing food consumption (Richardson 1973). Pharmacological stimulation of the amygdala with adrenergic agonists increases food intake and presumably hunger, but reduces water intake. On the other hand, cholinergic stimulation increases water intake but reduces food intake (Grossman 1964). Amygdalectomized rats emit consummatory responses to inappropriate objects and fail to avoid novel foods that would normally be treated as dangerous by intact animals (Rolls and Rolls 1973). Furthermore, stimulation of the amygdala can be used as an aversive stimulus, equivalent to foot shock, when an animal is tested in an escape task (Richardson 1973).

3.7.1 The role of the amygdala in CTA

A number of experiments have assessed the anatomical and functional involvement of the amygdala during toxiphobic conditioning processes. Large permanent lesions in the amygdala disrupt conditioned flavour aversions induced by a variety of agents serving as a US such as amphetamine, lithium chloride, and X-rays (McGowan *et al.* 1972; Grupp *et al.* 1976; Elkins 1980; Bermúdez-Rattoni *et al.* 1986). Furthermore, the amygdala has also been electrically stimulated during acquisition and extinction of toxiphobic conditioning. Stimulation of the amygdala before application of the CS resulted in stronger taste aversion in stimulated rats than in non-stimulated controls. However, if the animal received stimulation at the onset of toxicosis, taste aversion conditioning was disrupted (Phillips and Le Piane 1980). The same results were also obtained by Kesner *et al.* (1975) when they applied disruptive electrical stimulation after application of apomorphine used as an US. Furthermore, Buresova *et al.* (1979) have shown that licking of conditioned aversive flavours inhibits unit activity in the amygdala, as well as in the ventromedial hypothalamus and insular cortex of the rat. From these results, it is clear that the limbic system, and particularly the amygdala, is involved in the associative process of different learning tasks including the visceral associations established during toxiphobic conditioning.

It seems that the amygdala is highly involved in learning and memory processes implementing CTA. In recent years, this role of the amygdala has been debated due to the use of different experimental approaches. As mentioned above the central nucleus seems to play an important role in the acquisition and retention of aversive conditioned reactions, i.e. passive and active shock avoidance. However, differential effects on CTA were reported when specific nuclei of the amygdala were destroyed. Lesions of basolateral nuclei produced disruption of taste aversions (Nachman and Ashe 1974; Yamamoto *et al.* 1995*a*) but lesions of central amygdaloid nuclei did not (Galaverna *et al.* 1993; Yamamoto *et al.* 1995*a*). The debate concentrates on the role of the basolateral nucleus in the acquisition and retention of CTA (Yamamoto *et al.* 1994*b*). In a number of papers it has been established that lesions of the amygdala impair inhibitory avoidance learning (Dunn and Everitt 1988; Cahill and McGaugh 1990; Bermúdez-Rattoni and McGaugh 1991). However, CTA learning appears to be blocked only by large electrolytic ablations of the amygdala but not by excitotoxic lesions, which spare fibres passing through the amygdala (Dunn and Everitt 1988; Chambers 1990; Bermúdez-Rattoni and McGaugh 1991). Bermúdez-Rattoni *et al.* (1983) showed that reversible lesions induced by acute application of novocaine (a local anaesthetic) into the amygdala disrupted the acquisition of taste-potentiated odour aversions, but left taste aversions unaltered. Fitzgerald and Burton (1983) suggested that destruction of the fibres of passage and not the destruction of the basolateral nucleus itself is responsible for the observed CTA deficits. This question was directly addressed by Chambers (1990) who made either *N*-methyl D-aspartate (NMDA) lesions or large electrolytic lesions in the basolateral amygdala. The results showed that electrolytic lesions attenuated both acquisition and retention of CTA, whereas NMDA lesions affected neither CTA acquisition nor retention.

This contrasts with a recent finding that ibotenic acid induced lesions aimed at the basolateral nucleus of the amygdala have profound disruptive effects on both acquisition and retention of CTA, and that combined lesions of the insular cortex and amygdala produce still stronger CTA disruption (Yamamoto and Fujimoto 1991; Yamamoto *et al.* 1995*a*). These contradictory results can be explained by the different experimental procedures employed, i.e. by the different excitotoxic drugs employed. A recent series of papers demonstrated a differential effect of different excitotoxic lesions of the nucleus basalis magnocellularis (NBM) on amygdaloid and cortical cholinergic markers (Boegman *et al.* 1992; Mallet *et al.* 1995). Other authors have also found that ibotenic acid lesions of the NBM produced greater impairments in the acquisition of a water maze task than quisqualic acid lesions of the same nucleus, even though quisqualic acid produced a larger decrease in cortical ChAT activity (Connor *et al.* 1991). Another possible source of different results is the saccharin concentration employed by Yamamoto and Fujimoto (1991). These authors used a higher (0.2%) saccharin concentration than the other (0.1%). The stronger saccharin concentration could produce odours that in turn could be associated with the taste, suggesting that the effects seen are due to the disruption of taste-potentiated odour aversion. To examine this possibility Yamamoto *et al.* (1994) recently habituated rats to the smell of saccharin, by presenting it several days before making ibotenic acid lesions of the amygdala. In spite of this procedure, the lesions still caused significant disruption of CTA. The above findings do not exclude the possibility that amygdala lesions may actually produce disruption of taste-potentiated odour aversion. More careful studies are necessary to clarify these differences, by controlling the odours and the excitotoxic drugs employed.

3.7.2 Amygdala and taste-potentiated odour aversions

Flavours are composed primarily of odours and tastes which may differentially contribute to the formation of conditioned aversions to flavours (García *et al.* 1985; Bermúdez-Rattoni *et al.* 1986; see Chapter 5). Thus, taste is an excellent conditioning stimulus for delayed illness but odour is not. When odour and taste are combined into a compound stimulus and followed by illness, the odour becomes even more aversive than taste when each is tested separately in extinction trials. Odour aversions appear to be potentiated by taste, as if a sensory and-gate controlled by taste entered odours into food-related memory mechanisms (Palmerino *et al.* 1980; García *et al.* 1985). Several researchers have studied the neural mechanisms involved in the convergence of taste and odour stimuli and their associations with delayed visceral inputs. The amygdala receives gustatory, visceral, and olfactory afferents (White 1965; Norgren 1974) from the olfactory bulb and olfactory cortex. It appears to be a very suitable site for the interaction of odour with taste and toxicosis because it has all three modalities necessary for the mediation of taste-potentiated odour–illness conditioning.

In a series of experiments, Bermúdez-Rattoni and co-workers proposed that the amygdala and other parts of the limbic system may perform the gating of odour into the taste-potentiated aversions (Bermúdez-Rattoni *et al.* 1983, 1986, 1987*a*). In the first experiment, large lesions were used since previous studies indicated that large areas of the amygdala might be involved (McGowan *et al.* 1972; Grupp *et al.* 1976; Bermúdez-Rattoni *et al.* 1983). Therefore, rats with extensive electrolytic lesions including the lateral, dorsolateral, basal, and medial nuclei of the amygdala were compared with a group receiving sham lesions. These groups of rats were divided into subgroups, with odour, taste, or the odour-taste compound as CS, and all of them received LiCl 30 min later; the procedure was the same as described above. Unlike the sham operated rats the lesioned animals acquired neither odour, nor taste, nor taste-potentiated odour aversions. It was concluded that the integrity of the amygdala is essential for odour and taste aversion learning. However, in previous studies we found that reversible inactivation of the amygdala by novocaine disrupted only odour but not taste aversion. This suggests the possibility of discrete involvement of some nuclei rather than of the whole amygdala (Bermúdez-Rattoni *et al.* 1986; Fig. 3.3).

In the second experiment smaller lesions directed at specific nuclei of the amygdala complex were tested by conditioning aversions to the flavour compound and to its odour–taste components. Rats were given sham operations or small electrolytic lesions aimed at the basal amygdala including the lateral and basolateral, medial nuclei, central amygdala nucleus. Following postoperative recovery, each group received conditioning to an odour-taste compound followed by delayed illness. Almond odour and saccharin were the CSs while the US was LiCl. After conditioning, the tests with odour or taste alone showed that all groups displayed strong taste and odour aversions, except the group that had sustained lateral amygdaloid lesions, which displayed a significant disruption of odour aversion learning (Bermúdez-Rattoni *et al.* 1986). These results were confirmed later by Hatfield *et al.* (1992) who found that lesions induced by injection of neurotoxins into the basolateral nucleus of the amygdala produced similar results as described above, i.e. no deficits in taste or odour aversion learning but a significant deficit in taste-potentiated odour aversion learning. In addition, it has been postulated that NMDA receptors in the basolateral nucleus may play an important role in the taste-potentiated odour aversions, since intracerebroventricular injections of an NMDA antagonist D-2-amino-5-phosphonovalerate (D-APV) selectively blocked the taste potentiation of odour aversions (Willner *et al.* 1992; Hatfield and Gallagher 1995). Recently, it has also been found that depletion of dopaminergic terminals in the amygdala complex produces a strong disruption of taste-potentiated odour aversions but leaves taste aversions unaltered (Fernández-Ruíz *et al.* 1993, 1994). One of the roles of the amygdala in toxiphobic conditioning appears to be the association of odours presented together with taste with visceral consequences and implementation of the taste-potentiated odour aversion learning.

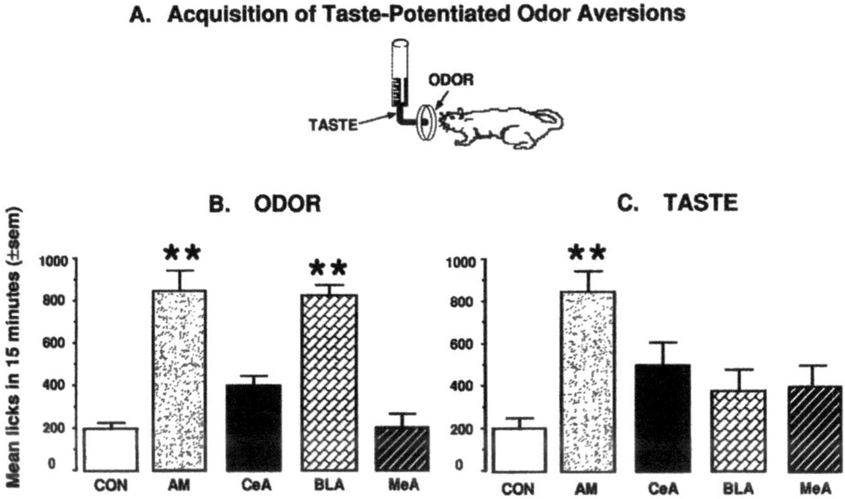

Fig. 3.3 Effects of large or small amygdala lesions on taste-potentiated odour aversion learning. (A) During the acquisition trial, presentation of a compound of taste (0.1% sodium saccharin dissolved in water) and odour (0.2 ml of almond extract on an odour disc) was followed by intragastric injection of 0.15 M LiCl (190 mg/kg) 30 min later. (B) Water consumption (expressed in lick rate) in the presence of odour alone following large amygdaloid lesions (AM), small lesions aimed at the central (CeA), basolateral (BLA), or medial (MeA) nuclei of the amygdala, or sham (CON) operations. (C) Consumption of saccharin solution following large (AM) or small lesions (CeA, BLA, MeA) of amygdala nuclei. Large amygdala lesions (AM) disrupted both taste and odour aversions. The small basolateral amygdala lesions (BLA) disrupted odour but not taste aversions. Neither central (CeA) nor medial (MeA) small lesions affected the development of odour and taste aversions. $^{**}P < 0.01$. Modified from Bermúdez-Rattoni *et al.* (1986).

3.8 Hippocampus

Investigation of the role of the hippocampus in taste aversion learning yielded contradictory results. Several studies have shown that lesions of the hippocampus have no effect on formation of CTA (Murphy and Brown 1974; Miller *et al.* 1975; Thomka and Brown 1975; Reilly *et al.* 1993*b*), whereas other researchers report a significant disruption of CTAs by hippocampal lesions (Best and Orr 1973; Krane *et al.* 1976; Nonneman and Curtis 1978; Miller *et al.* 1986). However, almost all of these researchers showed that hippocampal lesions disrupted the neophobic response to CSs.

Yamamoto *et al.* (1995*a*) showed that lesions of the whole hippocampus had no effect on acquisition, but attenuated the neophobic response to the saccharin CS, and showed a significant disruptive effect on CTA retrieval (Fig. 3.21). Separate lesions of the dorsal or the ventral parts of the hippocampus had no effect on neophobic responses, acquisition or retention of CTA (Fig. 3.2G, H). These results are consistent with those of other researchers. For example, Murphy and Brown (1974) showed that large aspiration of the hippocampus did not affect the speed of acquisition, strength, or resistance to extinction of CTA to sucrose paired with LiCl illness. Miller *et al.* (1975) reported that extensive aspiration lesions of the hippocampus did not affect the acquisition or strength of aversion to saccharin paired with cyclophosphamide, although the lesioned rats extinguished the aversion more quickly than control rats. Reilly *et al.* (1993*b*) showed that ibotenic acid lesions of the hippocampus had essentially no effect on acquisition of CTAs, but significantly impaired maze learning. These results suggest that hippocampal neurons do not play an important role in the acquisition of CTA or in the association of CS with US, but are concerned with previous experience classifying the stimulus as novel or familiar, as safe or dangerous. In this respect, it is of interest to note that Gallo and Candido (1995*a,b*) showed that dorsal hippocampal lesions impaired blocking (see 2.2.9) but not latent inhibition (see 2.2.7) of taste aversion learning in rats.

As already mentioned, some conflicting results of the limbic lesion studies are probably due to different combinations of tastes and flavours used to induce CTA. A number of experiments comparing the involvement of the hippocampus in CTA and passive avoidance learning lead to the conclusion that it is easier to disrupt the latter than the former (Gaston 1978). That is, several studies have failed to find deficits in the acquisition of taste aversion after hippocampal lesions (Miller *et al.* 1971, 1986; Murphy and Brown 1974; Bermúdez-Rattoni *et al.* 1987*a*). Further, Kesner *et al.* (1975) report that neither post-CS nor post-US stimulation of the hippocampus had any disruptive effect on CTA learning. Nevertheless, when irradiation rather than a poisonous drug was used as the toxic agent, rats with large hippocampal lesions failed to acquire saccharin aversion (Miller *et al.* 1971). Therefore, it is possible that in some of the above-mentioned experiments, odours present during the acquisition may produce, with different flavours, taste-potentiated odour aversions. Thus, amygdaloid or hippocampal lesions might produce their effect by blocking the association of odours and not necessarily of tastes with the visceral consequences of food consumption.

3.9 Entorhinal cortex

The entorhinal cortex (EC) has been ascribed an important role in mediating access of cortical sensory information to the hippocampal formation (Room and Groenewegen 1986), and may function as an integrating system related to modulation of cognition and memory. Norgren and Grill (1976) demonstrated by using a tract tracing method that there are corticofugal efferents from the insular cortex to the EC. Yamamoto *et al.* (1995*a*) showed that EC lesions resulted in reduced neophobia to novel taste CS. However, another finding that essentially no disruptive effects on CTA were observed after EC lesions (Fig. 3.21) suggests that corticoento-rhinal hippocampal pathways are not involved in formation of CTA.

3.10 Prefrontal cortex

The prefrontal cortex (PFC), which receives projections from the medio-dorsal thalamic nucleus, has mutual interconnections with the limbic structures, hypothalamus, and sensory cortical regions. Nauta (1971) has assumed that PFC constitutes the neural substrate for integration of interoceptive and exteroceptive information, thus mediating adaptive behaviour. To test this suggestion, Divac *et al.* (1975) ablated by suction or electrolytically either of the two regions in the frontal cortex of the rats, i.e. the mediodorsal cortex (mdPFC; Fig. 3.1) and the dorsal bank of the rhinal sulcus (sPFC; Fig. 3.1), in addition to the anteromedial part of the neostriatum to see the effects on the formation of taste aversion. However, they could not show any disruptive effects on the acquisition of CTA to saccharin which had been paired with i.p. injection of LiCl. They concluded that the prefrontal cortex in the rat is not necessary for association of external (taste) and internal (visceral) cues. Recently, however, Hernadi *et al.* (1995) showed deficits of CTA acquisition and retention after microlesions of mediodorsal PFC by iontophoretic application of kainic acid or 6-hydroxydopamine in the rat. They suggest that the PFC involving catecholaminergic mechanisms is a structure of great importance in memory storage and retrieval of learned taste information. Nevertheless, these two contrasting results require further studies on the role of the PFC in taste aversion learning as well as in regulation of feeding and integration of sensory information.

3.11 Insular cortex

The anatomical connections of the insular cortex (IC) clearly suggest that this brain region may play a role in integration, and possibly in storage of visceral information. Thalamic nuclei send projections to the IC located along the rhinal sulcus in the rat. The insular cortex in the rat is defined as an area spanning from the lateral frontal cortex to the perirhinal cortex in the rostrocaudal direction and from the ventral edge of somatomotor cortex to the piriform cortex in the dorsoventral direction (Saper 1982*a*; Zilles 1985). Furthermore, the IC has been described as the *visceral cortex* because it receives taste and visceral information from the thalamus and it is known to be involved in visceral reactions and stress (van der Kooy *et al.* 1984). The first reports of investigating the electrophysiological responses of the IC to taste stimulation of the tongue were published by Ganchrow and Erickson (1972) and Yamamoto and Kawamura (1972). Later, the gustatory area of the cortex receiving afferent connections from the thalamic taste area, the ventromedial thalamus, was confirmed by electrophysiological methods (Kosar *et al.* 1986). The gustatory neocortex occupies a 1 mm wide and 3 mm long region forming only 15% of the total IC area, and its boundaries do not extend beyond +3.0 mm rostrally and 0.0 mm caudally from bregma (Benjamin and Akert 1959; Yamamoto *et al.* 1980; Kosar *et al.* 1986). Furthermore, the gustatory neocortex is primarily located in the dysgranular insular cortex (Yamamoto 1984; Cechetto and Saper 1987; Ogawa *et al.* 1990), and the viscerosensory cortex in the granular insular cortex caudodorsal to the gustatory neocortex (Cechetto and Saper 1987; Ito 1992). The anatomical connections of the IC suggest that the IC plays an integral role in the mediation of visceral reactions. Moreover, it has been suggested that the IC is a site of convergence of limbic input with primary sensory inputs that is not seen within any other sensory area of the cortex (Krushel and van der Kooy 1988).

The IC connections that may be important for mnemonic processing include those with the limbic system, the amygdala, the dorsomedial nucleus of the thalamus, and the prefrontal cortex (Krushel and van der Kooy 1988). As mentioned above, the insular cortex and the central nucleus of the amygdala are functionally reciprocally interconnected (Yamamoto *et al.* 1984; Kapp *et al.* 1985; Lasiter and Glanzman 1985; Pascoe and Kapp 1987; Escobar *et al.* 1989). There is a direct projection between the amygdala and the IC via the internal capsule (Norgren and Wolf 1975; Kiefer 1985). Functional connections between the amygdala and the insular cortex have been demonstrated in single-unit studies showing that the insular cortex exerts influence upon neural activity within the central nucleus of the amygdala (Yamamoto *et al.* 1984; Pascoe and Kapp 1987). Apparently, this cortical projection may enable integration of perceptual and cognitive information with appropriate patterns of autonomic activity (Pascoe and Kapp 1987). Cortical projections to the amygdala, including those from the IC, have been suggested to convey processed cognitive information to be integrated with emotional and motivational processes (Pascoe and Kapp 1987).

The first experiments on the role of the insular cortex in cognitive processing of taste stimuli started at the beginning of the 1970s (Braun *et al.* 1972). Later, several studies confirmed that the IC is involved in mediation of the associative aspects of taste responding, but not in the hedonic responses to taste. That is, rats lacking the IC are impaired in acquiring and retaining taste aversion. Several experiments in different laboratories have demonstrated that IC lesions made either before or after CTA acquisition impair taste aversion learning or retrieval (e.g. Braun *et al.* 1972, 1981, 1982; Nachman and Ashe 1974; Yamamoto *et al.* 1980; Aggleton *et al.* 1981). Because IC lesions do not produce obvious deficits in gustatory or gastrointestinal sensitivity, it has been suggested that IC lesions impair the mnemonic representation of tastes and of their post-gastrointestinal consequences (malaise) (Nachman and Ashe 1974; Braun *et al.* 1982; Lasiter 1982). Furthermore, it has been demonstrated that the hedonic responses of lesioned IC rats appear to be normal; like normal rats, IC lesioned animals prefer sucrose, as well as low concentrations of sodium chloride, over water and reject quinine and acid solutions. This is not surprising because taste responsiveness remains intact even in decerebrated rats (Braun *et al.* 1982; Kiefer 1985).

Despite the fact that the IC lesioned animals show consistent deficits in the acquisition and retention of learned taste habits, these animals are able to detect taste stimuli normally (Kiefer and Braun 1979; Kiefer *et al.* 1984). However, it is important to note that disruption of CTA learning is relative. In an elegant series of experiments Kiefer and Orr (1992) showed that the failure of rats with insular cortex lesions to respond aversively to a taste paired with illness does not reflect a complete inability to generate aversive responses In fact, these authors showed that rats with IC lesions can produce aversive responses to strong concentrations of different solutions. The aversive responses were measured by a technique developed by Grill and Norgren (1978*a,b*) which makes it possible to assess the natural responses to tastes, i.e. disgust or unpalatability is reflected by stereotyped responses such as gaping, fluid expulsion, head shaking, and forelimb flailing in response to different fluids (see Grill 1985). The IC lesioned rats were able to show disgust to a strong HCl solution by gapes and passive drip of the taste solution; they did not show as much head shaking or forelimb flailing as the controls (Kiefer and Orr 1992). The authors interpreted these results as showing that the IC lesioned animals are unable to produce an hedonic shift of the taste which has been followed by illness, but retain their ability to avoid unpleasant natural tastes.

Yamamoto *et al.* (1995*a*) reported that lesions of the IC attenuated the acquisition and retention of CTA, but did not completely disrupt CTA formation (Fig. 3.2C). This indicates that the IC is not indispensable for CTA acquisition but is more important for CTA retrieval (Yamamoto *et al.* 1980*a*; Braun *et al.* 1981). It is conceivable that this result may be due to a compensatory effect. The fact that combined lesions of the IC and amygdala completely disrupt CTA acquisition (Yamamoto *et al.* 1995*a*) suggests that the IC could support CTA formation when the amygdala has been eliminated, and that this role can be taken over by the amygdala after IC damage.

Several studies tried to differentiate the topography of IC lesions by their effects on CTA. Hankins *et al.* (1974) found that very large anterior cortical lesions produced deficits in both flavour–illness and noise–shock conditioning but that large posterior cortical lesions disrupted noise–shock conditioning without affecting taste aversion learning. Lasiter (1982) and Nerad *et al.* (1996) confirmed that anterior lesions of IC including granular and agranular insular cortices surrounding the middle cerebral artery block CTA, and more posterior lesions including external capsule, claustrum, endopiriform nucleus, perirhinal cortex, and caudal parts of the caudate nucleus did not. This report also showed by neuroanatomical tracing techniques that the posterior lesions destroying the amygdalocortical input did not disrupt acquisition or retention of CTA. Further, it has been shown that lesions of the neocortex surrounding the anterior insula do not produce deficits in the acquisition of CTA, and that intracortical association projections appear to play a minor role in the formation and retrieval of taste–illness memories (Lasiter 1982). Therefore, it seems that a relatively specific portion of the rat's neocortex corresponding to the anterior insular cortex is involved in the acquisition and retention of learned taste aversions.

3.11.1 Insular cortex and taste-potentiated odour aversions

It is hypothesized that the insular cortex is also crucial for taste potentiation of odour stimuli. Antidromic responses in the olfactory bulb produced by stimulation of the piriform cortex indicate that there are projections from the olfactory bulb to the piriform cortex (Allen 1941). The piriform cortex sends direct axonal projections to the agranular insular cortex and to the terminal projection field in the orbital frontal neocortex (OFN), corresponding to the sPFC in the rat (Haberly and Price 1978). Lasiter and Glanzman (1985) used retrograde transport of horseradish peroxidase (HRP) to demonstrate that the primary gustatory neocortical projection field receives axonal projections from the PBN. Ablation of this region disrupts flavour–illness conditioning (Braun *et al.* 1972; Hankins *et al.* 1974). Topographically, OFN and IC lie close together in the lateral frontal brain region (Braun *et al.* 1985). Furthermore, Lasiter and Glanzman (1982) have described abundant connections between the OFN and IC. This suggests that olfactory and gustatory information may converge, at least in part, to the IC and that this convergence contributes to the sensations described by humans as 'flavours'.

To determine the role of the insular cortex in taste-potentiation of odour, intact rats and rats lacking the IC were given either a taste, an odour, or an odour–taste compound followed by intragastric intubation of LiCl (Kiefer *et al.* 1982). On the first exposure to the stimuli, rats lacking gustatory neocortex failed to display normal neophobia for the odour–taste compound. After two acquisition trials, separate test trials with the odour and taste components revealed the following effects of the lesions: taste conditioning was disrupted completely by IC ablations. The ablations affected neither odour conditioning nor potentiation of an odour aversion by taste. Although operated rats trained with the odour–taste compound formed no aversion to the taste component, these rats still displayed potentiated aversions to the odour component. Apparently, integrity of the insular cortex is not necessary for potentiation of odour by taste: the 'and-gate' function must be carried out elsewhere.

A series of related experiments by Kiefer *et al.* (1986, 1987) reported that rats lacking IC or olfactory bulbs showed normal acquisition of learned alcohol aversions, but extinguished more rapidly than intact controls (Kiefer *et al.* 1986). When the two types of ablation were combined, i.e. olfactory bulbectomies and IC lesion, little or no aversion learning was evident (Kiefer *et al.* 1987). These experiments clearly show that bulbectomized and IC-lesioned animals use different cues to learn the alcohol–illness associations. In another experiment the authors compared either IC-lesioned or bulbectomized animals with controls for aversion generalization to single taste–odour stimuli and to compound taste–odour stimuli following alcohol aversion training. The results showed that animals with IC but not with olfactory bulb lesions failed to show aversion generalization to any taste solution. It was concluded that animals with either type of ablation acquired the alcohol aversion using different cues: rats lacking IC acquired the alcohol aversion using primarily odour cues, and rats lacking olfactory bulbs using gustatory cues (Kiefer *et al.* 1987). A recent experiment demonstrated that catecholaminergic depletion of the insular cortex produced disruption of taste aversion, but not of taste-potentiated odour aversion in the same animals. Catecholaminergic depletion of the amygdala disrupted taste-potentiated odour aversion, but left development of taste aversions unaltered (Fernández-Ruíz *et al.* 1993).

The above results indicate that IC damage by either aspiration, electrolytic, or chemical lesions has profound disruptive effects on the acquisition of CTA. Experimental designs allowing good control of the odour and taste components of the flavour compound clearly showed that IC-lesioned animals lost their ability to learn taste but not odour aversions. When the same lesions were made in the amygdala, amygdalectomized animals lost their ability to acquire odour aversions and taste-potentiated odour aversions. It is possible to conclude that amygdala contributes to the integration of taste-potentiated odour aversion by gating the odours to the taste followed by illness whereas insular cortex only participates in the acquisition and retention of CTA.

4 Functional ablation studies of CTA

Interpretation of lesion studies reviewed in Chapter 3 is complicated by the inherent ambiguities of the ablation method. CTA learning leads to formation of appropriate LTMs, the existence of which can only be revealed by successful retrieval. Most lesion studies assume that acquisition and retrieval share the same neural mechanisms and are, therefore, simultaneously disrupted by the lesion. This is obviously not always the case as documented by various instances of anterograde amnesia. Thus lesion of the temporal cortex in the patient H.M. disrupted formation of long-term memories but did not interfere with retrieval of memories formed before the amnesic intervention (Milner 1972). Examination of the possible dissociation of acquisition and retrieval mechanisms by pre-acquisition and post-acquisition lesions was recommended by Thompson (1983) but was only rarely used in CTA experiments. An anterograde CTA amnesia, i.e. the failure to acquire new CTAs combined with good retrieval of the already learned CTAs, was reported by Schwartz and Teitelbaum (1974). The more common finding that CTA is disrupted by either pre-acquisition or post-acquisition lesions shows that the lesion prevented retrieval but tells nothing about its effect on acquisition. It is conceivable that new CTA memories are formed in the lesioned brain but that they cannot be retrieved. Unfortunately, in the absence of retrieval there is no way how to read the CTA file. Sudden blindness does not prevent the recall of visual information—the blind person can still describe the shapes and colours of familiar objects, but fails in a recognition test. Because CTA retrieval can only be tested by recognition in a single modality, interruption of the gustatory pathway disrupts both acquisition and retrieval. More complicated results are obtained in cases when acquisition and retrieval are implemented by different circuits, sharing the same gustatory input but connected to different outputs. A lesion blocking the CTA output circuit prevents retrieval without interfering with the acquisition circuit—but this possibility cannot be verified by the lesion technique alone. A more understandable result is obtained in experiments in which a post-acquisition lesion disrupts retrieval of previously acquired CTAs but does not prevent acquisition of the same or another CTA. In this case, the lesion has probably eliminated the particular CTA engram but did not preclude its possible duplication by the still available parallel circuits. The above possibilities are summarized in Table 4.1.

Most of the above uncertainties are due to the fact that the destroyed part of the brain may be important only for a short-lasting stage of the memory-forming process and is either not necessary at all or is used in a different context later. Consequences of permanent destruction indicate that the lesioned structure is important for the process studied but do not indicate which particular phase of the process has been affected. Resolving the spatial and temporal dynamics of the plastic changes underlying learning and memory by answering not only the 'where' but also the 'when' question requires the use of 'functional ablation procedures', the reversibility of which allows the brain region eliminated during the early stage of the process to be intact again during a later stage, particularly during the retrieval test.

Table 4.1 Effect of brain lesions on CTA retrieval in naive and overtrained animals

Lesion before training	Lesion after training	Interpretation
No retrieval	No retrieval	Retrieval failure with possible acquisition failure
No retrieval	Retrieval	Acquisition failure, independent retrieval mechanisms
Retrieval	No retrieval	Retrieval failure, compensated by parallel acquisition/retrieval mechanisms

4.1 Gross inactivation procedures

The first reports showing differential sensitivity of different stages of CTA to disruption appeared in the early 1970s, when Berger (1970) demonstrated that deep pentobarbital anaesthesia induced shortly after saccharin drinking and maintained during several hours of LiCl poisoning does not disrupt CTA formation. This result was confirmed by Roll and Smith (1972), Rozin and Ree (1972), Buresova and Bures (1977), and Rabin and Rabin (1984), and extended to other general anaesthetics, e.g. ketamine (Welzl *et al.* 1990) and rompun (Bermúdez-Rattoni *et al.* 1988). While the above studies prove that the GSTM–visceral US association proceeds under deep anaesthesia, it is more difficult to demonstrate the effect of anaesthesia on the initial processing of the gustatory stimulus. This is possible with subanaesthetic dosages of ketamine (25 mg/kg) which do not impair drinking. As shown by Welzl *et al.* (1990) saccharin ingested under such conditions cannot be associated with subsequent LiCl poisoning. This result was replicated by Aguado *et al.* (1994), who pointed out, however, the possible contribution of state-dependent learning to the CTA retrieval failure. They also reported the possibility to overcome the ketamine-induced CTA blockade by repeated pairing of the ketamine-influenced CS with LiCl. Strong CTA appeared after the second and third pairings. This indicates that the subanaesthetic dose of ketamine weakened but did not totally eliminate the GSTM which could be revealed by prolonged training.

Neither spontaneous nor forced feeding can be used in deeply anaesthetized animals but the taste stimulus can be applied in this case by the systemic administration of high concentrations of the taste substance. The taste buds in the oral cavity can be stimulated by the substance present in the blood after i.v. (Bradley and Mistretta 1971) or i.p. (Baum *et al.* 1974) injection. Buresova and Bures (1977) have shown that i.p. injection of saccharin (2% sodium saccharin solution, 1% body weight) followed by LiCl poisoning elicits significant CTA in intact rats but fails in rats anaesthetized with pentobarbital 15 min before injection of saccharin. The effectiveness of vascular taste as the CS was analysed by Bellingham and Lloyd (1987) who showed that the intensity of CTA increased with the length of the time interval between the i.p. injections of saccharin and LiCl and culminated at 2 h. This CS–US interval was employed by Bures and Buresova (1989) in an experiment comparing the effect of pentobarbital anaesthesia applied at various intervals before and after saccharin administration. No CTA developed (Fig. 4.1) when saccharin was injected into anaesthetized rats or rats recovering from anaesthesia. CTA was also prevented by pentobarbital injected up to 30 min after saccharin application. Anaesthesia induced later (1 h after saccharin) did not prevent formation of the GSTM and the GSTM–US association which produced CTA of the same strength as in unanaesthetized animals. These results suggest that the i.p. injected saccharin is detected only after a delay of about 30 min. Low efficiency of CTA acquisition with short CS–US intervals may be due to the fact that CS perception does not precede but follows the US. Such reverse conditioning is further confounded by the 'medicine effect', i.e. by positive labelling of the taste stimuli coinciding with the recovery from poisoning. These considerations do not affect, however, the main finding, clearly showing that even subanaesthetic dosages of pentobarbital block GSTM formation. However, once such trace has been formed, its association with the visceral signals of poisoning is not prevented even by deep anaesthesia.

A comatose state resembling general anaesthesia can be induced without drugs by purely physical treatment, by reducing body temperature to and maintaining it at a definite level. Using this technique, Ionescu and Buresova (1977*b*) showed that rats cooled 15 min after saccharin drinking to 20°C colonic temperature form the LiCl-based CTA with the same efficiency as intact animals. On the contrary, saccharin taste applied by i.p. injection to hypothermic rats could not be associated with subsequent poisoning, probably because GSTM formation was prevented. Since hypothermic rats can be force-fed, Buresova and Bures (1981) extended the above results by

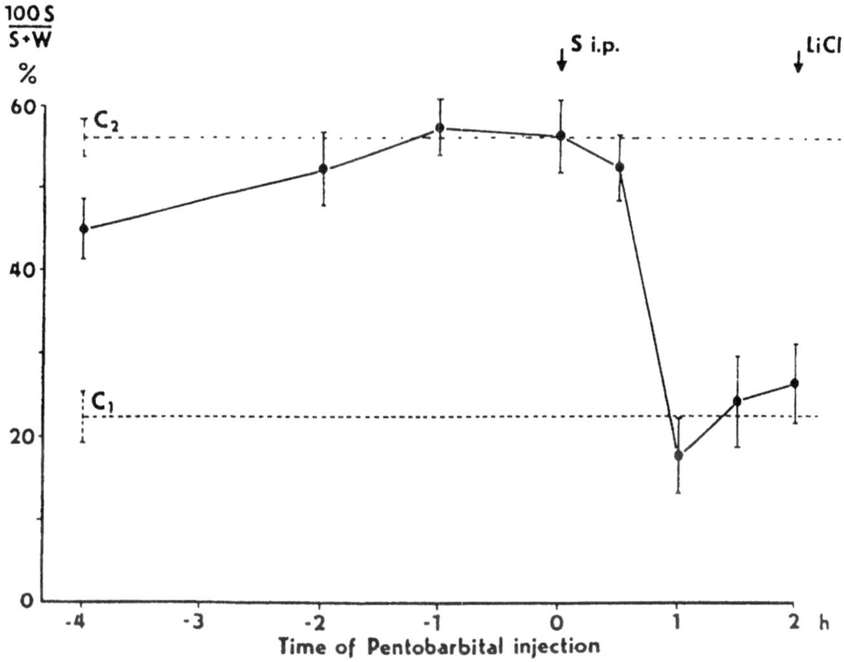

Fig. 4.1 The effect of pentobarbital (50 mg/kg) applied during CTA acquisition on the mean (±SEM) saccharin (S)–water (W) preference (ordinate 100S/[S+W]) during the two-choice retention test. Groups of eight rats. Pentobarbital was injected at various intervals (abscissa, hours) before (negative values) or after (positive values) intraperitoneal injection of saccharin (S i.p. arrow) followed 2 h later by lithium chloride (0.15 M, 2% body weight, LiCl arrow). The horizontal dotted lines correspond to saccharin preference in unanaesthetized poisoned controls (C_1) and non-poisoned controls (C_2). From Bures and Buresova (1989).

estimating the exact body temperature blocking GSTM formation. Saccharin (0.1%) force-fed to hypothermic rats did or did not elicit CTA at temperatures above or below 32°C, respectively (Fig. 4.2). This result is in good agreement with other behavioural findings indicating that a similar degree of hypothermia interferes with extinction of passive avoidance reaction in rats (Buresova *et al.* 1964). The abrupt onset of the GSTM blockade coincides with the hypothermia-induced inactivation of the reticular formation (Battista 1967).

4.2 Local inactivation procedures

Anaesthesia and hypothermia make it possible to block a definite phase of CTA acquisition but the locus of the interference is not well known. Better

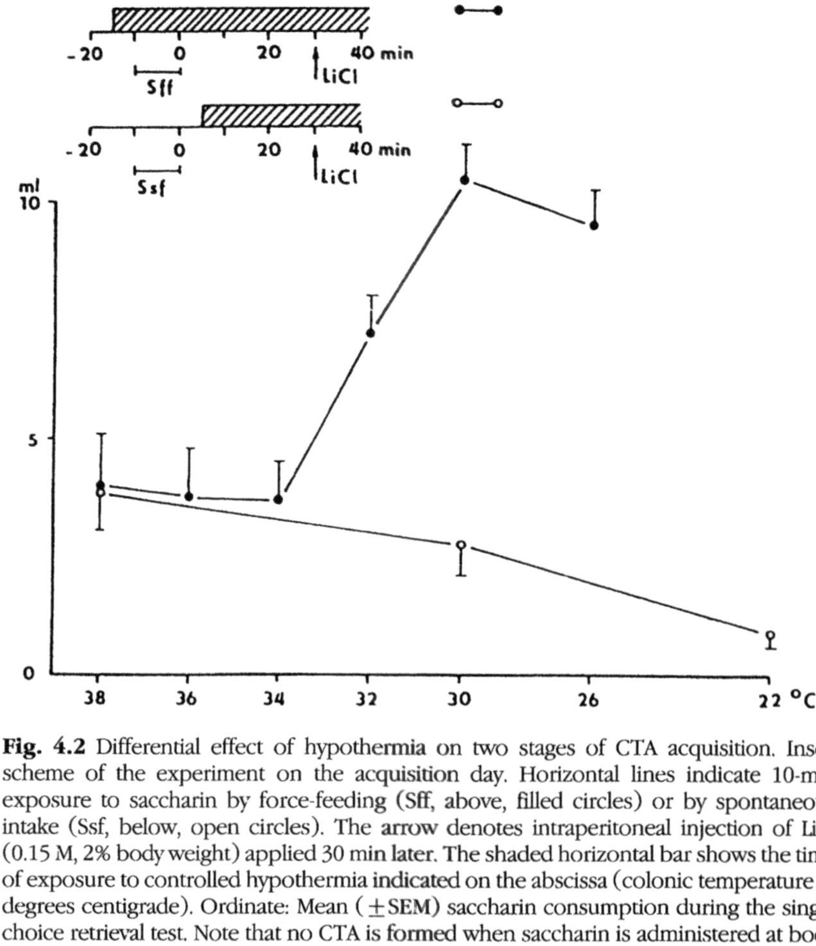

Fig. 4.2 Differential effect of hypothermia on two stages of CTA acquisition. Inset: scheme of the experiment on the acquisition day. Horizontal lines indicate 10-min exposure to saccharin by force-feeding (Sff, above, filled circles) or by spontaneous intake (Ssf, below, open circles). The arrow denotes intraperitoneal injection of LiCl (0.15 M, 2% body weight) applied 30 min later. The shaded horizontal bar shows the time of exposure to controlled hypothermia indicated on the abscissa (colonic temperature in degrees centigrade). Ordinate: Mean (±SEM) saccharin consumption during the single choice retrieval test. Note that no CTA is formed when saccharin is administered at body temperatures below 32°C, but that much deeper post-ingestion hypothermia does not decrease efficiency of poisoning.

spatial resolution can be achieved by inactivation of circumscribed brain regions presumably important for CTA learning.

4.2.1 Functional decortication by spreading depression

The first studies of this type (Davis and Bures 1972; Buresova and Bures 1973, 1974, 1975) concentrated on the role of the cerebral cortex in CTA acquisition and retrieval. Functional decortication by repeated waves of Leao's (1944) cortical spreading depression was used to disrupt for several hours EEG activity of unanaesthetized rats. Spreading depression elicited in both hemispheres by application of filter papers soaked with 25% KCl on the parietal cortex exposed by trephine openings 4 mm in diameter blocked cortical placing reactions and active avoidance reactions (Bures *et al.* 1974). It prevented formation of CTA to gustatory stimuli applied by force-feeding during functional decortication but did not interfere with CTA acquisition when applied in the CS–US interval or during the LiCl poisoning (Fig. 4.3A). Since spreading depression remains limited to the cerebral cortex and only rarely enters the amygdala and basal ganglia (Bures *et al.* 1974) the above results suggested that the cerebral cortex plays a critical role in the initial processing of the CS and in the GSTM formation. On the other hand, the GSTM is not localized in the cortex, because functional decortication did not disrupt its association with the US, but only accelerated its decay over the several hours long CS–US interval (Buresova and Bures 1974).

Functional decortication also interferes with CTA retrieval. A rat with strong CTA to saccharin refuses to swallow saccharin solution introduced into its mouth and reacts with typical aversive response (Grill and Norgren 1978*b*). Bilateral functional decortication blocks spontaneous drinking but the rat drinks the force-fed aversive fluid as eagerly as water (Brozek *et al.* 1974). It is obvious that the functional decortication suppressed the active aversive response protecting the animal against new poisoning. The failure of this defence reaction does not imply that the exposure to the gustatory stimulus remains undetected. New presentation of saccharin to the intact animal shows that the previous saccharin administration to the functionally decorticated (Buresova and Bures 1975) rats has significantly decreased CTA, probably by an extinction-like process. The extinction takes place even when saccharin administration to functionally decorticated rats is followed by poisoning.

4.2.2 Pharmacological blockade of the insular cortex

In an attempt to specify the critical cortical area inactivation was restricted to the gustatory insular cortex which was blocked by bilateral intracortical injection of 10 ng of tetrodotoxin (TTX) in 1 μl of saline (Gallo *et al.* 1992). Unlike functional decortication, elimination of gustatory cortex alone did not interfere with spontaneous drinking of saccharin but completely disrupted CTA formed by subsequent administration of LiCl (Fig. 4.3B). TTX applied after saccharin drinking and overlapping with the effect of LiCl poisoning did not prevent CTA formation (Gallo *et al.* 1992; Roldan and Bures 1994).

TTX blocks the voltage-dependent sodium channels and interferes, therefore, with the activity of somatodendritic membranes and axons. From this point of view it mimics surgical ablation of the treated brain volume. A more subtle intervention would leave neural activity of the treated area

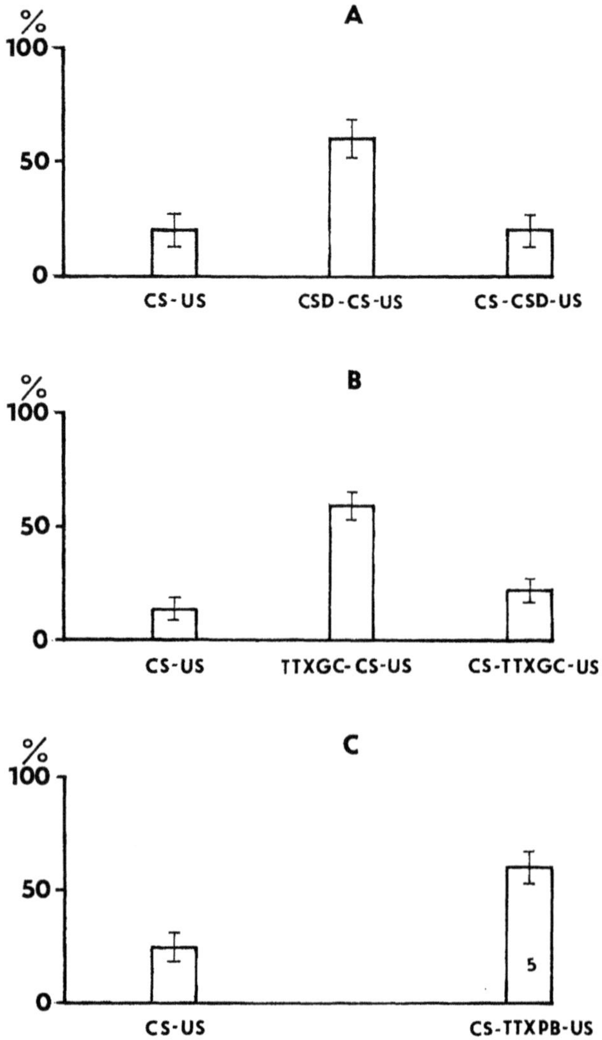

Fig. 4.3 The effect of functional blockade of the cerebral cortex or parabrachial nucleus on CTA acquisition. Ordinate: saccharin preference during the retrieval test. Abscissa: sequence of stimuli and treatments employed during CTA acquisition. CS, conditioned taste stimulus; US, unconditioned toxic stimulus; CSD, cortical spreading depression; TTX GC or TTX PB, blockade of gustatory cortex or of parabrachial nucleus by tetrodotoxin (10 ng). (A) Functional decortication by repeated waves of CSD blocks CTA acquisition when elicited before but not when elicited after application of the CS (Buresova and Bures 1973). (B) CTA acquisition is disrupted by pre-CS but not by post-CS blockade of gustatory cortex by TTX (Gallo *et al.* 1992). (C) CTA acquisition is prevented by post-CS injection of TTX into the parabrachial nucleus (Ivanova and Bures 1990*a*).

unaffected but would disrupt its participation in plastic processes presumably implementing formation of new memories. Since inactivation of protein synthesis impairs memory consolidation (Davis and Squire 1984), Rosenblum *et al.* (1993) injected 100 μg of anisomycin in 1 μl of saline into the gustatory cortices of both hemispheres. Anisomycin applied 15 min before saccharin drinking and followed 2 h later by LiCl prevented CTA. Anisomycin injected immediately after LiCl administration did not prevent CTA. Anisomycin applied before the retrieval test did not interfere with CTA readout. The anisomycin blockade of CTA was strictly site dependent: injection to a point 2 mm more medial was ineffective.

The results of the above experiments confirm the role of the gustatory cortex in CTA learning but show at the same time that it is restricted to a narrow time window corresponding to the presentation of the gustatory stimulus. The fact that the blockade induced immediately after drinking does not interfere with GSTM formation and persistence suggests that this trace is not stored in the gustatory cortex but at some extra-insular, probably subcortical site, where it may persist without cortical support. Spreading depression and TTX inactivations also resemble the lesion effect on CTA retrieval: the rats drink the CS fluid without recognizing its learned aversiveness. It must be pointed out, however, that the functional lesion elicits a more pronounced disruption than a comparable surgical ablation. According to Kiefer and Orr (1992) rats without a gustatory cortex are unable to learn CTA but can avoid the dangerous taste by using alternative food selection strategies. Still more striking is the effect of anisomycin which leaves retrieval intact. This indicates that anisomycin does not disrupt neural activity but mainly interferes with its modifications implementing learning. Another interesting difference between the spreading depression and anisomycin is that the former does not prevent CTA extinction (Buresova and Bures 1975) whereas the latter interferes with formation of gustatory memories implementing latent inhibition learning (Rosenblum *et al.* 1993). It is conceivable that neural circuits engaged in CTA learning are different from those used for formation of neutral or positive memories. Further research is required to find their storage site and the circuits bypassing the gustatory cortex.

4.2.3 Pharmacological blockade of extracortical centres

Functional ablation procedures were also used to examine the contribution of other forebrain and midbrain centres to CTA learning. TTX blockade of the amygdala was investigated by Gallo *et al.* (1992) and Roldan and Bures (1994). Bilateral injection of 3 ng TTX into both amygdalae before saccharin drinking did not interfere with CTA acquisition. On the other hand, TTX injected into the amygdala immediately after 15 min of saccharin drinking and 30 min before LiCl injection caused significant attenuation of CTA. Further experiments analysed the effect of TTX blockade of the amygdala on the GSTM–LiCl association in rats anaesthetized immediately after saccharin drinking with pentobarbital. Bilateral injection of 10 ng TTX was performed in the stereotaxic apparatus and followed by the LiCl injection. Whereas TTX injection into the amygdala disrupted CTA completely, similarly timed TTX injection into the gustatory cortex, hippocampus, bed nucleus of stria terminalis, lateral hypothalamus, and ventral thalamus was without effect. CTA disruption was observed only when TTX inactivation of amygdala overlapped with the LiCl effect, i.e. when the TTX injection was applied 1.5 h but not 6 h after administration of the US. This suggests that the functional ablation of the amygdala impairs the association between the GSTM and the poisoning induced signals but that it does not interfere with the consolidation of the GLTM.

Most important results were obtained with the TTX blockade of the parabrachial nucleus, a mesencephalic relay centre of the gustatory and visceral pathways, the damage of which interferes with CTA retrieval and prevents CTA acquisition (Di Lorenzo 1988*a*; Flynn *et al.* 1991; Spector *et al.* 1992; Yamamoto 1993). Because large PBN lesions interrupt both the CS and the US pathways, it is difficult to decide whether the effect is due to disruption of the input/output circuits or to interference with the GLTM storage. The following four possibilities were discussed by Spector *et al.* (1992): (a) blockade of the conditioned aversive response (cessation of licking, attempts to remove the aversive fluid from the mouth); (b) impairment of gustatory signal processing; (c) impairment of visceral signal processing; (d) failure in the integration of taste signals and visceral signals and in the formation of the CTA trace.

Some of these explanations can be directly rejected when TTX blockade of PBN is used instead of a lesion. Ivanova and Bures (1990*a*,*b*) showed that CTA formation can be prevented when the PBN of rats which have been anaesthetized after saccharin drinking is inactivated by TTX for the duration of the LiCl poisoning (Fig. 4.3C). The CTA absence during the retention test performed two days later cannot be explained by impairment of aversive responding or by reduced taste sensitivity, because at that time the animals have fully recovered from the TTX effect and their gustatory stimulus processing and oromotor responsiveness were normal. PBN blockade induced shortly after LiCl injection could have interfered with the processing of the visceral signals of poisoning and, thus, disrupted the GSTM–US association. Although this mechanism cannot be excluded, it is not the only one, because CTA was also disrupted by TTX applied 24 h after saccharin–LiCl pairing. This indicates that the TTX inactivation of PBN interferes with processes taking place after cessation of the CS and US signals, i.e. with events related to the formation and consolidation of the GLTM. This conclusion is consonant with the finding that the PBN blockade was ineffective when TTX was injected after the second of two conditioning trials or 8 days after a single CS–US pairing, that is at a time when the engram has already reached a more advanced stage of consolidation and/or has perhaps migrated to a storage site outside the PBN. Supporting evidence was obtained in lesion studies. Grigson *et al.* (1997*b*) reported that rats receiving electrophysiologically guided PBN lesions after three CTA conditioning trials retained the acquired CTA but were not able to learn a new CTA.

Comparison of the effect of post-trial TTX injections into the gustatory cortex, amygdala, and PBN shows that no retrograde amnesia is obtained in the cortex, a brief amnesia (corresponding to the duration of poisoning) in the amygdala, and a prolonged amnesia from the PBN, which seems to be indispensable for the integration of the gustatory CS with the visceral US and for the subsequent storage of the CTA engram. The convergence of the gustatory and visceral inputs in the PBN region (Hermann and Rogers 1985) is not a sufficient condition for CTA learning, however, as was demonstrated by the inability of chronically decerebrated rats to acquire CTA (Grill and Norgren 1978*c*). The gustatory cortex is necessary for the initial processing of the taste stimulus and the amygdala participates in the mediation of the visceral US. It is conceivable that during simultaneous activation of the PBN and gustatory cortex, corticofugal feedback may cause priming of the corresponding PBN neurons into a state representing the GSTM. For several hours, the primed PBN neurons are prepared to be associated with the visceral signals of poisoning, the effectiveness of which may be enhanced by feedback from the amygdala. Existence of corticofugal connections to PBN was demonstrated by experiments showing that novocaine blockade of the gustatory cortex modulates PBN responses to taste stimuli (Di Lorenzo 1990) and that taste responsive PBN neurons can be activated by electrical stimulation of the gustatory cortex (Di Lorenzo and Monroe 1992).

Evidence supporting continuation of the consolidation process in the PBN was recently obtained by Krivanek (1996), who measured the activity of protein kinase C (PKC) in the cytosol and particulate fractions of PBN at various time intervals after CTA acquisition. An increase of cytosol activity was found 24 h (+18%) and 48 h (+40%) after a single saccharin–LiCl pairing. Twelve hours and 5 days after CTA acquisition PKC activity did not differ from the control levels. The above PKC increase was not influenced by saccharin presentation immediately preceding the anaesthesia used for killing the animals before preparation of the PBN tissue and reflects, therefore, a slow autonomous process initiated by the CS–US pairing, but proceeding independently of the ongoing brain activities. In this respect the PKC activity increase clearly differs from the *c-fos* reactivity changes appearing in various PBN subnuclei after CTA retrieval (Yamamoto 1993).

The delayed activation of PKC synthesis was not elicited by injection of LiCl alone (Krivanek 1993) and was not observed outside PBN, e.g. in the occipital cortex. CTA acquisition was disrupted by PBN injection of polymyxin B, but not by the more specific PKC inhibitors like H-7 or staurosporin (Bielavska and Krivanek 1994). Further research is required to confirm the putative relationship between the PKC activation and CTA consolidation at the PBN level.

4.3 Lateralization of the CTA circuits

Surgical and functional lesions can be used not only for mapping the brain areas critical for a particular type of memory in the whole brain but also for confining the engram formation to a definite brain region. A particularly important paradigm introduced by Sperry (1961) is interhemisperic transfer of learning in the surgically (Doty and Ringo 1994) or functionally (Bures *et al.* 1974) split brain. In a typical experiment of the latter type the rat learns a task during blockade of one hemicortex by repeated waves of Leao's (1944) spreading depression. Retrieval is tested during functional decortication of the trained hemisphere while the originally depressed hemisphere is intact. Two extreme results can be obtained. (1) The task learned with one hemisphere cannot be retrieved with the contralateral hemisphere when the functional block has been switched. This result suggests that the memory trace formed during the acquisition trial is confined to the brain regions blocked by the functional ablation procedure during the retrieval test. (2) The memory acquired during unilateral functional decortication can be retrieved with the contralateral hemisphere as efficiently as with the same hemisphere. This indicates that the engram has also been formed outside the brain region blocked during retrieval, i.e. in the case of functional decortication in subcortical structures.

The reversible split brain technique was used to assess the lateralization of GSTM formation by Lehr and Nachman (1973), who allowed thirsty rats with unilateral spreading depression in the right hemisphere to drink 0.15 M LiCl for 10 min. During a retention test performed with functional ablation of the same hemisphere, LiCl consumption dropped by 80% but almost no CTA was observed during blockade of the opposite (untrained) hemicortex. Lateralization was less pronounced after two acquisition sessions with the same hemisphere. No significant lateralization of CTA acquired during functional hemidecortication was observed by Buresova and Bures (1973), who used 0.1% saccharin as the CS and i.p. injection of LiCl as the US. During retrieval of the unilaterally acquired CTA, saccharin intake was decreased by 80 and 65% with functional decortication of the same and the opposite hemispheres, respectively. The results indicate that unilaterally acquired CTAs may be lateralized in some cases but that it is more common that the memory readout can also be accomplished through the untrained hemisphere. This may indicate either that the memory trace is bilaterally represented or that a lateralized trace can be accessed through both hemispheres.

Better CTA lateralization can be achieved by extending the functional split-brain approach to subcortical structures. Gallo and Bures (1991) combined functional hemidecortication with unilateral PBN blockade. Rats drinking saccharin during spreading depression in the right hemisphere and receiving unilateral injection of TTX before i.p. injection of LiCl formed CTA when both interventions were applied to the same hemisphere but not when applied to different hemispheres (Fig. 4.4A). The CTA retrieval was tested with intact brain in this case. Rats drinking saccharin with intact brain and receiving unilateral PBN blockade during LiCl administration learned a weak CTA, the retrieval of which was similarly disrupted by spreading depression elicited in either ipsilateral or contralateral hemicortex. These results suggest that CTA acquisition requires the interaction of the cerebral cortex with the PBN through ipsilateral pathways (Saper 1982*a*,*b*; Shipley and Sanders 1982) implementing the corticofugal priming of the PBN neurons.

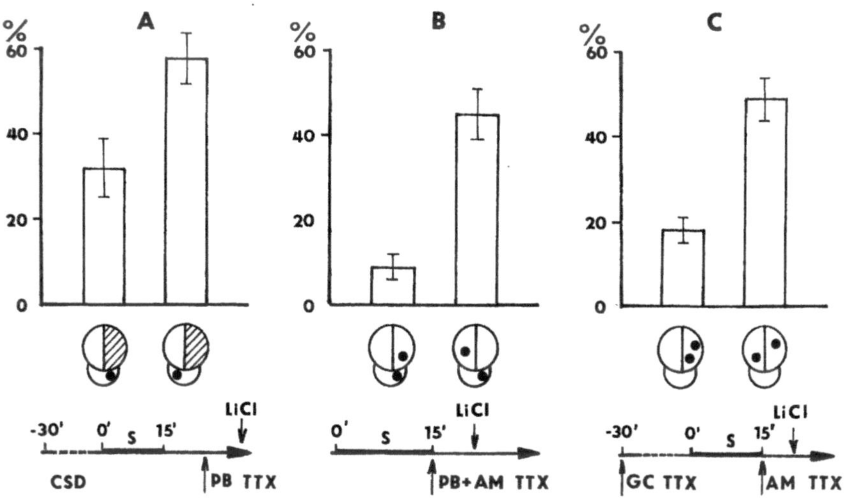

Fig. 4.4 Lateralization of the neural circuits mediating CTA learning. (A) Combined functional hemidecortication (CSD, shaded hemisphere) and unilateral blockade of the parabrachial nucleus by TTX (PB TTX, black dot). (B) Combined TTX blockade of the parabrachial nucleus and amygdala (PB + AM TTX, black dots). (C) Combined TTX blockade of the gustatory cortex (GC TTX, black dot) and amygdala (AM TTX, black dot). The columns indicate mean (±SEM) percentage of saccharin (S) intake from total fluid intake during the retrieval test. Schemes of cerebral hemispheres and brainstem below the columns illustrate conditions during CTA acquisition. The horizontal arrow indicates timing (min) of the interventions (CSD, TTX, LiCl) relative to the onset of saccharin drinking (S).

The above experiments were further extended by Bielavska and Roldan (1996) who examined the effect of simultaneous unilateral functional ablations of the insular cortex, amygdala, and PBN in the same or opposite hemispheres on CTA acquisition. Rats with implanted guiding tubes received 30 min before saccharin drinking an injection of 10 ng TTX into the gustatory cortex of one hemisphere and after saccharin presentation another TTX injection into the same or opposite amygdala. Retention tested 2 days later with the intact brain showed the absence of CTA in the group with contralateral inactivation of the gustatory cortex and amygdala (Fig. 4.4C). Similar results were obtained in another series of experiments in which post-drinking TTX injections into the amygdala and PBN disrupted CTA when applied into different hemispheres but not into the same hemisphere (Fig. 4.4B). Strong CTA developed when one hemisphere was intact during CTA acquisition.

The importance of ipsilateral connections was also demonstrated for CTA retrieval. Rats which had learned CTA during the functional ablation of one PBN and had, therefore, a lateralized GLTM in the intact hemisphere, could retrieve CTA when TTX inactivation was applied to the gustatory cortex and/or the amygdala of the naive but not of the trained hemisphere. Particularly disruptive was combined elimination of both the amygdala and gustatory cortex of the trained hemisphere. The importance of ipsilateral cortico-amygdalar connections was demonstrated earlier by Buresova (1978) who showed that CTA acquired by rats with unilateral amygdalectomy can be retrieved during functional decortication of the lesioned, but not of the intact, hemisphere.

Schafe *et al.* (1995*b*) have recently addressed lateralization of CTA circuits by surgical lesions. These authors exploited the earlier finding that exposure to the aversively conditioned taste induces in CTA-trained animals a significant increase of c-*fos*-like immunoreactivity (c-FLI) in the NTS (Swank and Bernstein 1994). This increase disappeared after CTA extinction (Houpt *et al.* 1994) and was not elicited at all when the oral cavity was flushed with aversive concentrations of quinine (Swank *et al.* 1995). Since CTA cannot be formed in chronically decerebrate rats (Grill and Norgren 1978*c*) the authors assumed that the c-FLI increase was induced in the NTS by a forebrain input. They tested this assumption in rats with unilateral brain transection at the level of the superior colliculus. The chronically hemi-decerebrate rats acquired after two saccharin–LiCl pairings almost as strong an aversion to saccharin as the intact controls. The finding that unilateral connections between forebrain and brainstem can support CTA acquisition and retrieval is in good agreement with similar conclusions based on unilateral functional ablations of the insular cortex, amygdala, and PBN in the same hemisphere (Gallo and Bures 1991; Bielavska and Roldan 1996). Schafe *et al.* (1995) found the c-FLI increase elicited by forced presentation of saccharin only in the NTS retaining normal neural connections with the forebrain. This unilateral expression of c-FLI excludes crossed forebrain–brainstem connections below the level of the cut. Histological analysis of the extent of the transection suggested that the pathways mediating the induction of c-FLI in the NTS do not run close to the midline but instead occupy a more lateral position.

5 Pharmacology of CTA

Since the discovery of CTA in the 1950s, many psychologists considered CTA as a variant of classical or Pavlovian conditioning, while others classified it as a special kind of learning, a model of biologically specific cue-to-consequence adaptive learning (see Chapter 1). In this model, animals learn to associate a novel taste, the conditioned stimulus (CS), with a malaise inducing stimulus, the unconditioned stimulus (US). However, as we will see later this concept has several drawbacks.

The first difficulty encountered when studying the drug effects on CTA starts with the definition of the CS and US. Traditionally, the conditioned stimulus is a stimulus neutral to the organism that by association with the US, which produces physiological changes, starts after several CS–US pairings to control the animal's responses. In CTA research, unlike in Pavlovian conditioning which requires several conditioning trials, a single pairing of both stimuli leads to association of the CS with the agent-induced malaise (the unconditioned response; UR). Consequently, the next exposure of the CS elicits a conditioned response (CR) similar to the UR, which triggers the avoidance response (García *et al.* 1974; García 1989). However, in CTA conditioning some stimuli could be simultaneously used as both CS and US. For instance, a solution of LiCl has both components: it has a salty taste when ingested and after several minutes of drinking starts to induce malaise by its effects on the intestinal tract. Conditioned taste aversion is observed on the next occasion when the salt is presented. To identify the physiological mechanism serving as the US, García and Ervin (1968) proposed that the malaise or the disruption of the 'milieu interne' was the US eliciting conditioning. Thus the internal consequences of the drug-induced or agent-induced illness are associated with the ingested foods. In this regard, several drugs, toxins, or agents can produce CTA at specific doses (see Table 5.1). Further, several authors have attempted to classify the CTA-producing drugs according to their pharmacological effects on neurotransmitters (see Table 5.2). Nevertheless, such classifications are still confusing and do not shed much light on the physiological mechanisms of drug-induced CTA.

5.1 The emetic effects of drugs

The first pharmacological classifications of CTA-inducing drugs were based upon their emetic effects. For example, many drugs that produce internal

Table 5.1 Toxins and agents that have been reported to be effective in inducing taste aversions. Modified from Riley and Tuck (1985).

Acetaldehyde (Brown *et al.* 1978)
Acrylamide (Anderson *et al.* 1982)
Adriamycin (Bernstein *et al.* 1980)
Aflatoxin B1 (Rappold *et al.* 1984)
Alloxan monohydrate (Brookshire *et al.* 1972)
Allyl sulfate (Maruniak *et al.* 1983)
Alpha-methyl-p-tyrosin (Carey & Goodall 1974)
Alpha naphthylthiourea (Rzoska, 1953)
Arsenic (Rzoska 1953)
Barium carbonate (Rzoska 1953)
Bombesin (Deutsch & Parsons 1981)
CI-628 (King & Cox 1976)
Cadmium chloride (Wellman *et al.* 1984)
Cannabichrome (Carcoran *et al.* 1974)
Cannabigerol (Corcoran *et al.* 1974)
Carbaryl (MacPhail 1982)
Chloral hydrate (Kallman *et al.* 1983)
Chlordiazepoxide (Goudie 1981)
Chlordimeform (MacPhail & Leander 1980)
Chlorphentermine (Booth *et al.* 1977)
Cholecystokinin (Deutsch & Hardy 1977)
Cobalt chloride (Wellman *et al.* 1984)
Cobra venom (Islam 1978)
Copper sulfate (Nachman & Hartley 1975)
Cortical spreading depression (Freedman & Ward 1976)
Cycloheximide (Bolas *et al.* 1979)
Cyclophosphamide (Ader *et al.* 1978)
d-amphetamine + chlorpromazine (Cappell & LeBlanc 1971)
Deltamethrin (MacPhail 1982)
1,2-dichloroethane (Kallman *et al.* 1983)
1,2-dichloroethylene (Kallman *et al.* 1983)
2-deoxy d glucose (Thompson & Zagon 1981)
Delta-8-THC (Corcoran *et al.* 1974)

Delta-9-THC (Amit *et al.* 1977)
Desipramine (Lorden & Nunn 1982)
Dichlorvos (see Riley & Tuck 1985)
DiethDichlorvos (see Riley & Tuck 1985)
Diisobutyrlapomorphine (D'Mello *et al.* 1981)
Diisoprophylfluorophosphate (see Riley & Tuck 1985)
d-isoproxerenol HC1 (Margules 1970)
d,l-cathinone (Foltin & Schuster 1981)
Emetine (Revusky & Gorry 1973)
Epinephrine (Caza *et al.* 1982)
Ergocornine hydrogen maleate (King & Cox 1976)
Ergocornine methane sulfonate (King & Cox 1976)
Ethanol (Cannon *et al.* 1977; Kiefer 1995)
Estrogen (Gustavson *et al.* 1989)
Ether (Vogel & Nathan 1975)
Ethylenedine citrate (Panksepp *et al.* 1977)
Ethylenediamine dihydrochloride (Panksepp *et al.* 1977)
Fenfluramine (Booth *et al.* 1977)
5-fluoroucil (Bernstein & Webster 1980)
Fluoroacetate (Howard *et al.* 1977)
Fluoxetine (Lorden & Nunn 1982)
Formalin (Woods *et al.* 1971)
Hashish (Corcoran 1973)
Histamine diphosphate (Levy *et al.* 1974)
Histidine free amino acid load (Booth & Simson 1974)
(-)-hydroxycitrate ethylenedime (Panksepp *et al.* 1977)
(-)-hydroxycitric acid (Panksepp *et al.* 1977)
Hypertonic procaine (Mineka *et al.* 1972)
Insulin (Domjan & Levy 1977)

Ipecac (Baker & Cannon 1979)
Irradiation (Garcia & Koelling 1967)
Isotonic procaine (Mineka *et al.* 1972)
Lead acetate (Dantzer 1980)
l-amphetamine (Carey & Goddall 1974)
Leydig tumors (Bernstein & Fenner 1983)
Lithium carbonate (see, Riley & Tuck 1985)
Lithium chloride (Archer *et al.* 1979)
Mercuric chloride (Klein *et al.* 1974)
Mescaline (Cappel & LeBlanc 1971)
Mestranol (Howard *et al.* 1968)
Mesurol (Gustavson *et al.* 1982)
Methiocarb (Mason & Reidinger 1983)
Methotrexate (Ader 1977)
Methyl bromide (Miyagawa 1982)
Methylmercury (Braun & Snyder 1973)
Methylnitrate (Clody & Vogel 1973)
Metrazol (Golus & McGee 1980)
Monosodium glutamate (Vogel & Nathan 1975)
n-butyraldoxime (Nachman *et al.* 1970)
Neutron exposue (Garcia & Kimeldorf 1960)
Nicotine (Etscorn 1980)
Nitrogen mustard (Bernstein & Webster 1980)
Oxythiamine (Seward & Greathouse 1973)
Paraquat (see Riley & Tuck 1985)
Parathion (see Riley & Tuck, 1985)
p-chloramphetamine (Goudie *et al.* 1976)
p-chloromethamphetamine (Booth *et al.* 1977)
p-chlorophenylaline (Nachman *et al.* 1970)
Phencylidine (Etscorn & Parson 1979)
Phenobarbital (Vogel & Nathan 1975)
p hydroxyamphetamine (Booth *et al.* 1977)
Physostigmine sulfate (Parker *et al.* 1982)
Pilocarbamide (Gay *et al.* 1975)
Plant homogenates (Yokel & Ogzewalla 1981)

Proferrin (Feinberg 1973)
Pseudo coriolis effect (Lamon *et al.* 1977)
Red squill (Nachman & Hartley 1975)
Rotation (Braun & McIntosh 1973)
Scopolamine methyl nitrate (Berger 1972)
Scorpion venon (Islam 1980)
Serum from irradiated donors (Garcia *et al.* 1967)
Sesame oil (Deutsch *et al.* 1976)
Skin secretions (Mason *et al.* 1982)
Smoke (Etscorn, 1980)
Sodium chloride + polyethylene glycol (Kutscher & Wright 1977)
Sodium fluoride (Ionescu & Buresova 1977)
Sodium fluoroacetate (Nachman & Hartley 1975)
Sodium iodacetate (Inoescu & Buresova 1977)
Strychnine sulfate (Howard *et al.* 1968)
Strychnine sulfate (Nachman & Hartley 1975)
1,2-trichloroethane (Kallman *et al.* 1983)
2,4,5-trichlorophenoxyacetic acid (Sjoden *et al.* 1979)
Thallium sulfate (Nachman & Hartley 1975)
Thiabandazole (Gustavson *et al.* 1983)
Thiamine deficiency (see Riley & Tuck 1985)
Trichloroethylene (Kallman *et al.* 1983)
Tricholoromethane (Laundauer *et al.* 1982)
Tricholoromethane (Kallman *et al.* 1982)
Triethyltin (MacPhail 1982)
Trimethyltin (MacPhail 1982)
Tryptophan -deficient diet (Treneer & Bernstein 1983)
Vasopressin (Ettenberg *et al.* 1983)
Viper venom (Islam *et al.* 1980)
Win 35.428 (D'Mello *et al.* 1979)

Table 5.2 Compounds affecting various transmitter systems and psychoactive drugs that produce a CTA. Modified from Gamzu *et al.* (1985)

Cholinergic

Agonist
Carbachol (Myers & DeCastro 1977)
Nicotine (Kumar *et al.* 1983; Parker 1995)
Physostigmine (Parker *et al.* 1982)

Antagonist
Scopolamine (Berger 1972)
Atrophine (Preston *et al.* 1984)
Methyl-Scopolamine (Berger *et al.* 1973)
Methyl-Atropine (Emmerick & Snowden 1976)
Mecamylamine (see Gamzu *et al.* 1985)

Catecholaminergic

Agonist
D-Amphetamine (Booth *et al.* 1977; Parker 1995)
L-Amphetamine (Booth *et al.* 1977)
Hydroxyamphetamine (Booth *et al.* 1977)
Methamphetamine (Booth *et al.* 1977; Parker 1995)
p-Clormethamphetamine (Booth *et al.* 1977)
Phenylpropanolamine (Wellman *et al.* 1981)
Cathinone (Foltin & Schuster 1982)
Cocaine (D'Mello *et al.* 1981)
Methylphenidate (Riley & Zellner *et al.* 1978)
Amitriptyline (Miller & Miller 1983)
Apomorphine (D'Mello *et al.* 1981)
Buproprion (Miller & Miller 1983)

Antagonist
Chlorpromazine (Berger 1972)
AMPT (Carey & Goodall 1974)

Opiate (mu)

Agonist
Morphine (Cappell *et al.* 1973; Parker 1995)
Methadone (Chipkin & Rosecrans 1978)

Antagonist
Naloxone (Stolerman *et al.* 1978; Frenk & Rogers 1979)
Naltrexone (Stolerman *et al.* 1978)
BC 2860 (Stolerman *et al.* 1978)
Mr 1452 (Stolerman *et al.* 1978)

Serotoninergic

Agonist
5-HT (See Gamzu *et al.* 1985)
5-HTP (Zabik & Roache 1983)
Fenfluramine (Booth & Simson 1973)
Fluoxetine (Lorden & Nunn 1982)
MK212 (See Gamzu *et al.* 1985)
Quipazine (See Gamzu *et al.* 1985)

Antagonist
PCPA (Nachman *et al.* 1970)

Psychoactive

Diazepam (Gamzu 1977)
Chlordiazepoxide (Gamzu 1977)
Flurazepam (Vogel & Nathan 1976)
Lorazepam (Berger 1972)

Anxiolytics/Hypnotics
Amobarbital (Vogel & Nathan 1975; 1976)
Hexobarbital (Vogel & Nathan 1976)
Pentobarbial (Vogel & Nathan 1976)
Methaqualone (Vogel & Nathan 1976)

Antidepressants

Amitriptyline (Miller & Miller 1983)
Buproprion (Miller & Miller 1983)

Anaesthetics

Ketamine (Etscom & Parson 1979)

malaise can also induce CTA. Thus, food aversions can develop in the course of medical treatments that induce emesis, e.g. in chemotherapy of alcoholism (Bernstein and Webster 1980; Logue 1985). CTA has been used to explain anorexia developing in cancer patients receiving chemotherapy and radiotherapy (Bernstein and Fenner 1983; Lazlo 1983). These treatments have been established as effective unconditioned stimuli in the acquisition of CTA (Bernstein 1985). The first studies taking into account the biochemical changes underlying CTA were initiated by García *et al.* (1967) who demonstrated that aversion-eliciting exposure to sublethal doses (10–30 r) of ionizing radiation is accompanied by release of toxic substances into the circulation. These toxins could be responsible for the emetic effects that mediate the feedback producing the acquired aversive behaviour (García 1989). In a later study, Levy *et al.* (1970) concluded on the basis of 'Sephadex' gel fractionation that the active substances obtained from muscle bath preparations contain common biogenic amines. By using several pharmacological blocking agents, they were able to identify bradykinin, serotonin, and mainly histamine as putative substances released during CTA acquisition (Levy *et al.* 1970).

Following this rationale, Lee and Green (1989) showed that the H_1 blocker chlorpheniramine blocks CTA but not analgesia and, conversely, that cimetidine, an H_2 blocker, disrupts analgesia but does not block acquisition of flavour aversions. The H_1 histaminergic receptors, neurons, and endings have been found in the brain areas involved in the pathways that contribute to the acquisition and maintenance of CTA (Lee and Green 1989). H_1 histaminergic receptors are typically found in the periphery, mainly in the smooth muscles. It is known that H_1 histaminergic terminals contribute to smooth muscle contraction (bronchial wall, blood vessels, and gut). Further, disruptive effects by H_1 blockers have also been reported when oestrogen was used as the US in CTA conditioning (Rice *et al.* 1987; Gustavson *et al.* 1989). In this regard, according to some reports, injections of histamine disrupt in rats the formation of taste aversions induced by LiCl and $CuSO_4$, and reduce CTA induced by subcutaneous injections of oestrogen in rats (Rice 1988). Finally, it should be pointed out that endogenous oestrogen has also been involved as a contributing factor to 'anorexia' in adolescent girls (Young 1991).

5.2 Toxicity effects in the formation of CTA

Other authors have tried to classify CTA-inducing drugs by their capability to produce vomiting or nausea. The remarkable work by Nachman *et al.* (1970) demonstrating that alcohol can produce CTA, suggested that CTA could be used for behavioural assays of drug toxicity. Many pharmacologists have used CTA in a battery of tests to assay the toxicity index of any particular drug (Gamzu *et al.* 1985). However, there are drugs or agents known to be toxic and or to produce illness which do not produce CTA. These findings were summarized by Riley and Tuck (1985) in a contingency table in which a toxic agent is categorized as a hit when it induces taste aversion and as a miss when it does not induce taste aversion; a false alarm is recorded when CTA is elicited by a compound with no evidence of its toxicity, and finally, correct identification applies to a non-toxic drug which does not produce reliable CTA.

The above toxicity interpretation of CTA is made questionable by misses for drugs that are known to be toxic but fail to induce taste aversions. Inclusion in the 'miss' category should eliminate other explanations, e.g. the drug dose, duration of drug action, delayed onset of the drug action, number of conditioning trials, etc. The best known CTA-inducing compound is LiCl. Whereas a low dose of LiCl applied in a single training trial does not induce CTA several pairings of the same taste with the same low dose of LiCl can produce CTA (Riley and Tuck 1985). Similarly, there are differences in the effective dosage utilized and the species-specific effects of the same drugs. In this regard, in a review of vomiting studies (King 1990), dogs were found to be more sensitive to different emetic agents than cats in 48% of reviewed cases, whereas cats showed more sensitivity than dogs to such agents in only 8% of the reports. Additionally, non-human primates and pigeons seem relatively insensitive to emetic stimuli, but the author pointed out that the latter species have not been tested as systematically and frequently as dogs, so those conclusions must be taken with some caution. It is possible that species-specific differences are due to the different connections of the vomiting centre. Thus radiation-induced emesis in the cat can be abolished by high spinal cordotomy plus vagotomy, but not by area postrema lesions, which do, however, abolish radiation or LiCl-induced emesis in dogs and monkeys (Fox *et al.* 1990; King 1990). Similarly, the dopaminergic neurotransmitter system is related to radiation-induced emesis in dogs but not in non-human primates or ferrets (King 1990).

5.3 Paradoxical effects of drug reinforcers

Gamzu *et al.* (1985) pointed out the paradoxical effect of drugs that can be operationally defined as positive reinforcers, since they can sustain self-administration, but can at the same time serve as aversive stimuli since they induce CTAs. As an example, let us take the study by van der Kooy *et al.* (1983) demonstrating that rats injected with apomorphine in a specific location, where they had drunk saccharin, avoided saccharin, but displayed preference for the location associated with the apomorphine injection. In the same vein, doses of benzodiazepines used to produce CTAs are equivalent to doses which in rats and monkeys release operant behaviour that has been suppressed by electric shock (Gamzu *et al.* 1985). Among drugs that can be self-administered and produce CTAs are morphine, phencyclidine, cocaine, amphetamine, ethanol, nicotine, barbiturates, etc. (Gamzu *et al.* 1985). There are a number of agents that are agonists or antagonists of various CTA-inducing neurotransmitters and psychoactive drugs (see Table 2). One of the clearly paradoxical effects is due to scopolamine, an anticholinergic agent traditionally used as an antiemetic drug that can, however, readily produce CTA. Conversely, it can attenuate aversion when injected before CTA conditioning trials in animals that are exposed to sucrose—LiCl pairing (Evenden *et al.* 1992). Hunt and Amit (1987) made an extensive review of such paradoxical effects and concluded that CTAs induced by drugs such as morphine are quantitatively distinct from CTAs induced by purely emetic agents, and that those differences can be detected by their pharmacological effects. Thus, the effects of purely emetic drugs are generally greater in magnitude and show less between-subject variability of taste aversion responses compared to the CTAs induced by self-administered drugs. Furthermore, the dose–response curve seems to be linear for the CTAs induced by pure emetics but curvilinear for CTAs elicited by self-administered drugs (Hunt and Amit 1987). Parker (1995) has recently reviewed over 10 years of research investigating the nature of the flavour–drug association produced by lithium and by rewarding drugs. The taste reactivity test developed by Grill and Norgren (1978*a,b*) and described in Chapter 2 was used to evaluate responses to intraoral infusion of the taste stimulus after CTA induced by reinforcing and emetic drugs. Conditioning made flavours paired with lithium chloride unpalatable; the animals displayed aversive responses like gaping, chin rubbing, etc. On the other hand, conditioning did not make flavours paired with rewarding drugs unpalatable. These results do not support the hypothesis that CTA produced by rewarding drugs is mediated by the side-effect of non-specific sickness (Parker 1995).

5.4 Endogenous peptides in the formation of CTA

Paradoxical effects have also been found in experiments investigating the role of endogenous peptides in the formation of taste aversion. Similar to drugs, these compounds have dual effects depending on the dose and physiological conditions of the animal. This is the case for cholecystokinin (CCK), a gut brain neuropeptide hormone which is believed to be a good candidate for a satiety factor. Thus, it has been demonstrated that animals secrete proportional amounts of CCK in response to a meal (116 ng CCK/ 1 ml of blood) within 30 min postprandially. Exogenous administration of CCK produces satiety when its concentration in the blood appears to be within physiological limits (165 ng/ml in rats). Under normal circumstances eating to the level of satiety produces a 'state of affairs' that the organism does not avoid but instead tends to attain or prolong by emitting operant behaviour. Since satiety in people is associated with warmth and satisfaction (Mueller and Hsiao 1978), it is logical to speculate that CCK is a powerful reinforcer.

Contrary to the above assumption, Deutsch and Hardy (1977) showed that CCK can be used as an aversive stimulus to produce bait shyness in rats.

They used 40 units per kg of the terminal octapeptide that had been shown to be an intermediate dose of CCK for eliciting food reduction. A two-bottle design was shown to be more sensitive for that aversive conditioning. Another gut peptide, bombesin, has also been reported to reduce intake in a two-bottle test of conditioned taste aversion (Deutsch and Parsons 1981). These results were confirmed by Smith *et al.* (1982), who reported that a single pairing of 2 μg/kg of the CCK octapeptide with a novel saccharin solution decreases subsequent saccharin preference. These results showed that externally administered CCK may produce some kind of malaise contrasting with the reinforcing properties of the CCK-induced satiety. It has also been reported in human subjects that injections of CCK at doses lower than those used to elicit CTA in animals produce nausea, headache, diarrhoea, and abdominal cramps (Sturdevant and Goetz 1976). Recently, it has also been demonstrated in humans that intravenous (i.v.) administration of CCK-8 produced nausea and emesis (de Montigny 1989). Furthermore, Swerdlow *et al.* (1983) found with a conditioned place-preference paradigm that a place that had been paired with CCK became aversive and a place paired with food was preferred. They concluded that decreased food consumption due to CCK is caused by malaise rather than by the satiety effect.

Some of the controversial results after CCK administrations could be explained by the peptide characteristics, i.e. salience, peak concentration, and duration after injections. In an attempt to investigate those possibilities, Ervin *et al.* (1995) repeated injections of small amounts of CCK-8 and compared the resulting CTA with that produced by a single large dose of the same peptide. Their results showed that distributed CCK administration produced significantly stronger CTA than a large single dose of CCK. The authors concluded that the strength of CCK-8-induced taste aversions depends upon the duration of peptide activity induced by repeated injections of small doses of the peptide (Ervin *et al.* 1995).

It has been postulated that taste and other oropharyngeal stimuli play an important role in the control of food intake (Mueller and Hsiao 1977, 1978). Thus, it is conceivable that CCK affects taste perception (and consequently food intake). Waldbilling and Callaghan (1980) reported that CCK-8 significantly reduces sucrose intake in non-deprived rats. Hsiao and Deupree (1983) found that CCK suppresses food rewarded, but not non-rewarded, bar pressing. On the other hand, BBS has suppressed both reinforced and non-reinforced lever pressing.

The idea that CCK can affect or modulate the hedonic properties of food was proposed by Gosnell and Hsiao (1984) who showed that CCK-8 decreases the number of licks emitted by satiated or *ad libitum* fed rats during short-term exposure to sucrose. The suppressive effects of CCK-8 were less expressed in hungry than in satiated animals. Integrated *chorda tympani* responses to stimulation of the tongue with sucrose or saline were recorded before and after i.v. infusions of CCK-8 or saline. The aim of the latter experiment was to demonstrate possible CCK modulation of the peripheral parts of the taste system. The results showed responses to sucrose and saline tended to increase after CCK infusions but not after saline infusions (Gosnell and Hsiao 1983). This study did not clearly indicate the effect of CCK on the modification of taste inputs at the level of taste receptors and fibres. Nevertheless, the hypothesis cannot be ruled out because the integrated response of the whole nerve can obscure the CCK related changes in single taste fibres forming the taste pathway.

5.5 Stress-related peptides in the formation of CTA

The mechanisms involved in responses to stress may also be involved in the formation of CTA. Thus, Smotherman *et al.* (1976) found that rats exposed to pairing of milk with LiCl showed elevated plasma levels of corticosteroids on the second presentation of milk. These results were replicated by Ader (1976) who produced a strong CTA by pairing saccharin with cyclophosphamide. In the group that showed the aversion, levels of corticosteroids, measured 3 or 6 days after acquisition, were significantly higher than in the appropriate control group. However, when rats were given a free choice to consume either water or the flavoured fluid previously paired with LiCl, reduced preference for the aversive taste was not accompanied by elevation of the plasma corticosterone level (Smotherman *et al.* 1976).

The role of pituitary adrenal activation mediating some forms of CTA has been demonstrated in a series of reports involving the pituitary adreno-cortico-tropic hormone (ACTH). Pituitary peptides, like ACTH, ACTH 4–10, and alpha-MSH (melanin-stimulating hormone), were shown to produce delayed extinction of appetitively motivated responses (Garraud *et al.* 1974). However, structural modification in the peptide sequence of ACTH 4–10 changes the effects on shock-motivated behaviour. Replacement of the phenylalanine residue in position 7 of the sequence 4–10 by its D-isomer (ACTH 4–10 7-D-Phe), facilitates extinction of active avoidance responses such as running or jumping (De Wied 1969; Garraud *et al.* 1974). On the other hand, the isomers 7-D-Phe and 7-L-Phe have the same effect on passive avoidance behaviour. Ritger and Popping (1975) studied the effects of alpha-MSH and of the two isomers of ACTH 4–10 in two different taste aversion procedures: forced extinction, when the target-flavour (sweet water) was the only fluid available to the animals during extinction, and the preference drinking test, when the animal could choose between tap water or sweet water during extinction. The results showed that both isomers of ACTH 4–10 and alpha-MSH delayed extinction in the preference test but not in the forced extinction procedure. It is noteworthy that '7LPhe' and '7DPhe' ACTH fragments had the same retarding effect on CTA extinction. Since these isomers of ACTH had opposite effects in animals trained to perform 'active' behaviours (De Wied 1969), their effect on taste aversion resembles that on passive avoidance. This means that the differential effects of the isomers depend on the nature of the behavioural response. CTA procedures can easily assess the possibility to induce aversion by exogenous injections of pituitary peptides alone. Since the organisms already have endogenous levels of these compounds at the time of the aversion testing, exogenous administration can lead to an overdose. Therefore, Smotherman and Levine (1978) tested the effects of ACTH and ACTH 4–10 on the preference drinking test in animals previously exposed to pairings of milk and LiCl. They found that ACTH has more pronounced aversive effects than ACTH 4–10; however, both peptides reliably accelerated extinction in comparison with the saline group.

In a second experiment the same authors paired milk with ACTH, corticosterone, or saline, administered 20 min after drinking. Neither of the substances tested had any aversive effects. ACTH given before three consecutive extinction sessions produced a reliable decrease of milk consumption during the second and third sessions. ACTH 4–10 had no effect during the three extinction days. On the other hand, a single pre-treatment with ACTH or ACTH 4–10 did not enhance neophobia although repeated injections of ACTH suppressed drinking. The rapidity with which the suppression occurred appeared to be inversely related to the familiarity with the substance.

The obvious question is: are the high levels of corticosterone found in CTA-trained rats necessary for conditioning? Attempts to answer this question have been made by measuring the levels of corticosteroids in adrenalectomized animals and comparing them with values found in normal animals (Frumkin 1975). Unfortunately, confounding variables may weaken the conclusions drawn from these data. For example, the lack of steroids caused by extirpation of the adrenal glands produces physiological changes leading to an increased need for salt which may affect conventional taste aversion procedures (Frumkin 1975). Ader *et al.* (1978) studied the effects of adrenalectomy on taste aversion conditioning but gave some low doses of dexamethasone and aldosterone (the hormone regulating salt transport in the kidney) to the operated animals in order to maintain the 'normal physiology of the animals'. Because both adrenalectomized and control groups showed reliable taste aversion after pairing of saccharin with cyclophosphamide, adrenal response is not critical for CTA acquisition and maintenance.

5.6 The CS–US delay

Another problem important for classification of the pharmacological aspects of drug-induced CTAs is the CS–US interval. As already mentioned, in the case of CTA it is not clear what is the optimum delay for the presentation of US and CS. In laboratory conditions, the association is established in a single trial by pairing the ingestion of the flavoured CS and the injection of the toxic drug that in turn produces malaise (García *et al.* 1974). In typical examples of conventional conditioning, optimal learning occurs when the CS precedes the US by 500 ms to 2 s (Domjan 1980; Abrams and Kandel 1988). In contrast, robust taste aversions can be acquired with taste–illness delays from 2 s to 12 h (García *et al.* 1982). However, not much is known about the minimum CS–US delay required for robust CTAs. Schafe *et al.* (1995*a*) investigated this issue by exposing the animals to intraorally delivered CS and to intravenously injected malaise-inducing agent. Parametric studies of the CS–US interval were made possible by presenting rats with saccharin and apomorphine injections at intervals of 1 s, 10 min, and 30 min. The results were paradoxical: as expected, the 10 and 30 min delays elicited robust CTAs, but the 1 s delay produced a very weak taste aversion (Schafe *et al.* 1995*a*). Other experiments in that direction have been made with the malaise-inducing agent presented up to 15 min before the novel taste (backward conditioning) (Yamamoto *et al.* 1994*c*). In this study significant aversions were reported, but were generally not as strong as those obtained with forward conditioning. Taken together, all these findings indicate that the onset of the US seems to be an important factor that has not yet been taken seriously into consideration.

5.7 Taste interaction with other stimuli

Additional characteristics of tastes should be taken into account, like their interaction with other stimuli participating in the formation of reliable CTAs. In general, it has been proposed that odour is processed as an external CS, resembling the audio-visual CS in terms of its associative affinities. Thus odour alone is a weak CS for a delayed toxic US, but it is an effective CS for immediate footshock US, paralleling the conditioning results obtained with the audio-visual CS (García *et al.* 1985).

Odour alone is a weak cue for delayed illness, but after odour has been presented in compound with taste before delayed illness, the odour component becomes a strong associative cue when tested alone during extinction. An experimental analysis (Rusiniak *et al.* 1979) revealed the following pattern: taste alone was a strong CS for a delayed toxic US. Conversely, odour alone was a weak CS for the same US. When taste and odour were presented as a compound, the odour component became a strong CS when tested alone under extinction conditions. This effect was termed *taste potentiation of odour* to indicate that a previously weak odour cue became a strong associative cue simply because it was presented in conjunction with a taste.

The potentiation of odour by taste, besides being relatively specific to illness conditioning, also depends upon the specific arrangement of stimuli at the time of conditioning (Holder *et al.* 1988). In a study by Palmerino *et al.* (1980), odour and/or taste combinations were presented in one pre-exposure trial and then paired with poison in a second acquisition trial. The following results were obtained: potentiation of the odour aversion was found only under those conditions when taste and odour were presented as a compound on the poison trial. Association of odour and taste on the pre-exposure trial (sensory preconditioning) did not weaken the final odour aversions. Previous research by Rusiniak *et al.* (1979) demonstrated that higher-order conditioning was also ineffective in flavour–poison conditioning.

Unlike odour, taste is specific to food ingestion. In a similar sense, gastric illness is typically related to ingestion of a tainted meal. Thus, gastrointestinal distress, because it is also food related, necessitates a specific search within taste memories of feeding. The retrieval and association of taste with illness are mediated by the convergence of neural mechanisms rather than by the classic association by contiguity of taste and illness. However, odour associated by contiguity with taste becomes a strong associative cue for the illness stimulus. Let us suppose that taste operates a gate through which contiguous odour gets access to the memory banks of the feeding system (García *et al.* 1982). Once the odour is indexed as 'food-related' it is no longer available for association with the peripheral foot-shock, therefore taste appears to block odour in flavour–shock conditioning. Furthermore, indexed odour operates as a taste, acquiring the parametric and selective properties of taste (Palmerino *et al.* 1980; Rusiniak *et al.* 1982). The associative mechanisms implementing the odour–taste connection do not apply to other external stimuli, used as CS. Thus in an experiment by Holder *et al.* (1988) noise paired with saccharin was 30 min later followed by an injection of LiCl. The results showed weak taste-potentiated noise–illness association, but the noise used was readily associated with shock when the latter was used as US.

5.8 The CS–US–FB model

Possible explanations of the drug effects on CTA and on taste-potentiated odour aversions require further investigation of each drug under a new scheme proposed recently by John García (1990). He argues that CTA can be conceptualized in a more comprehensive conditioning model based on a CS–US–FB (homeostatic feedback) format (Fig. 5.1). According to this model CS is an external signal, that can be detected by sensory organs like eyes, ears, or nose, whereas US represents stimuli contacting the organism, like an intraoral taste stimulus, and the homeostatic regulatory processes (FB) adjust the hedonic value of any specific US. As can be seen in Fig. 5.1, the taste stimulus (US) reaching the oral cavity starts to elicit the taste reactivity responses (UR elicited by the US). Then the FB regulatory processes are activated by specific stimuli (e.g. by the gastric effects of LiCl), that in turn alter the hedonic value of the US as a part of the more general defence system. The model should be taken into consideration because it includes the neural mechanisms involved in homeostatic regulation, particularly the specific neural structures and pathways that detect toxins, association mechanisms, and the homeostatic processes involved in conditioning (Ossenkop and Eckel 1995). Further study of drug effects using this model might improve our understanding of which drugs could change the FB and consequently affect the hedonic value of the US.

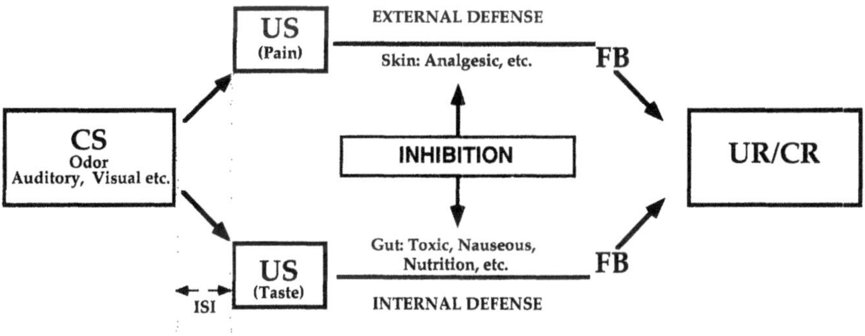

Fig. 5.1 Block diagram of the CS–US–FB conditioning paradigm, illustrating the external (skin) defence and the internal (gut) defence selective associations. CS, conditioned stimulus; US, unconditioned stimulus; FB, homeostatic feedback mechanisms; CR, conditioned response; UR, unconditioned response; ISI, interstimulus interval. Modified from García *et al.* (1989).

5.9 Neuropharmacology of flavour aversions

To investigate the neural mechanisms involved in the neural gating between odour and taste Bermúdez-Rattoni *et al.* (1983) performed a series of experiments, using bilaterally implanted cannulae for local application of selected drugs into different brain areas. In those experiments 'reversible' amygdala lesions were produced by local infusions of procaine, suppressing normal electrophysiological activity for about 2 h. In an initial experiment, four groups of rats were given a single exposure to an odour–taste (OT) compound. One group was a normal control, a second group was given procaine just before presentation of the compound OT, a third group was given procaine following OT exposure, and a fourth group received the procaine just before LiCl application. The administration of LiCl was given to all rats 30 min after exposure to the odour and taste. Following OT–illness, all groups displayed approximately equivalent taste aversions, reducing saccharin consumption. However, the group given procaine before the OT exposure displayed attenuated odour aversions. These data suggest that disruption of odour potentiation might have been produced by amygdaloid dysfunction just before the CS experience (Bermúdez-Rattoni *et al.* 1983).

The deficits in odour potentiation found in procaine-treated rats could be due to disrupted integration of odour and taste information. It is also possible that infusion of procaine into the amygdala disrupts simple odour detection or odour–illness associations. To test the possibility that odour perception was affected by procaine treatment, three groups of rats (normal, procaine pre-odour, and handled control) were given a single exposure to almond odour while drinking familiar tap water. After a 1-min exposure the rats were given a footshock (1 mA, 1 s). The results from the subsequent odour test indicated that all three groups showed significant suppression of consumption in the presence of the odour. These data indicate that procaine-treated rats were not anosmic, since they could develop normal associations between an odour and foot-shock; in fact, the treated rats exhibited superior odour–shock performance.

The third experiment tested whether amygdala dysfunction produced by procaine infusion would disrupt direct odour–illness conditioning induced by immediate toxin injection. Control rats (both saline infused and handled) and rats given procaine prior to odour exposure were conditioned with a single odour–illness trial using immediate administration of the poison so that direct odour–illness learning could be obtained. The results of the subsequent odour test indicated that pre-treatment with procaine disrupted conditioning.

These results suggest that the disruptions in potentiation of odour aversions described above can be attributed to a deficit in the formation of odour–illness associations, and that the amygdala dysfunctions may not have been related to odour–taste integration but rather to odour–illness associative learning. The effect was specific to toxiphobia conditioning because odour–shock learning was not affected by procaine treatment (Bermúdez-Rattoni *et al.* 1983). Recently, other investigators have been able to block the potentiated odour aversion with systemic, intraventricular or intracerebral (into the basolateral nucleus of the amygdala) injections of a competitive NMDA antagonist D-APV, without interfering with the expression of learning (Willner *et al.* 1992; Hatfield and Gallagher 1995). Neither of these treatments affected the development of CTA to taste alone or simple conditioning to an odour cue alone. These results suggest that the glutamatergic neurotransmission is highly involved in the gating of odour by taste in the flavour aversion conditioning. Recently, there has been a debate about the effects of NMDA-receptors antagonist in the formation of flavour memories. Thus Welzl *et al.* (1990) found that injections of ketamine, a non-competitive NMDA-receptor antagonist, before the presentation of a simple taste prevents the formation of taste aversions (Welzl *et al.* 1990). However, NMDA-receptor antagonists seem to affect the salience of the taste stimulus rather than the CTA formation (Aguado *et al.* 1994). Furthermore, other authors have found that competitive and non-competitive NMDA-receptor antagonists can readily induce CTA when injected after presentation of a new taste (Jackson and Sanger 1989; Myhal and Fleming 1990). These results indicate that NMDA-receptor-dependent plasticity mechanisms are involved in the taste-potentiated odour aversion but that their role in the formation of simple taste aversion remains unclear.

As explained in Chapter 3, the limbic system is highly involved in the formation of taste-potentiated odour aversion. Thus, animals with amygdala lesions are unable to acquire odour, taste, or odour and taste aversions. Furthermore, extensive damage to the amygdala can have devastating effects upon acquisition of both flavour–illness aversions and buzzer–shock avoidance tasks (McGowan *et al.* 1972; Grupp *et al.* 1976; Elkins 1980). Lesion studies (Chapter 3) do not fully agree with drug studies which affect primarily the odour–taste integration. The apparent discrepancy may be due to the fact that drugs produce a 'reversible' inactivation only during acquisition but not during retrieval. Obviously, permanent lesions affect associative, memorial, and motivational processes participating in both acquisition and retrieval.

Another limbic structure that has been involved in the integration of associative and cognitive processes is the dorsal hippocampus. In addition, the cholinergic activity of the hippocampus has been related to cognitive deficits induced by ageing and cholinergic deafferentation by septal lesions (Sinden *et al.* 1995). In a series of experiments, Bermúdez-Rattoni *et al.* (1987*a*) investigated the involvement of cholinergic neurotransmission in the regulation of taste–odour potentiation. Cholinergic agonists and antagonists were applied into the dorsal hippocampus before a taste–odour compound was presented to the rats. The results showed that the infusions markedly reduced aversion to the odour component but not to the taste component. Paradoxically, the agonists produced disruption of the odour aversion while the antagonist produced facilitation. It has been reported elsewhere that infusion of physostigmine or carbachol (cholinergic agonists) into the amygdala disrupted passive shock avoidance but not active shock avoidance. As shown in Chapter 8, cholinergic activity of the insular cortex is highly involved in the recovery of lost ability to acquire CTA after brain damage, as well as in the formation of taste aversion memories (Bermudez-Rattoni *et al.* 1995). Shimura *et al.* (1995) found that intraoral perfusion with CTA eliciting saccharin solution produced significant enhancement of cholinergic release from the insular cortex measured by *in vivo* microdialysis, similar to that elicited by perfusion with aversive bitter solutions like quinine.

Figure 5.1 can be improved by adding an inhibitory mechanism between the external (skin) defence system and the internal (gut) defence system. There is some evidence which indicates that passivity of the external system, induced by tranquillizers, facilitates taste aversion learning (Bermúdez-Rattoni *et al.* 1988). Further, it has been demonstrated that flavour aversion can be acquired under anaesthesia (Roll and Smith 1972). That is, the US–FB pairing changes the incentive value of the US, but the subjects may be 'unaware' of its associative causes. Since digestion and absorption are slow processes that follow a meal, the FB mechanisms operate while animals are asleep. If we further assume that the mutual inhibitory system, as well as the gating mechanism, lies within the limbic system, or is closely connected to it, then our paradoxical effect can be speculatively explained. For example, agonists exciting the inhibitory system would suppress the odour aversion as was shown above with infusion of cholinergic agonists into the amygdala and hippocampus. Functional or permanent inactivation of the amygdala would render the gate inoperative and reduce the odour aversion by other means (Bermudez-Rattoni *et al.* 1983, 1988; Ferry *et al.* 1995; Hatfield and Gallagher 1995). Destruction of the entire amygdaloid complex disrupts all aspects of odour, taste, and illness interaction (see Chapter 3). All this agrees with the above discussion. Furthermore, there is some evidence that cholinergic agonists infused into the amygdala also increase inhibition of external coping mechanisms, as was shown by Goodard (1969) and Todd and Kesner (1978). On the other hand, cholinergic antagonists infused into the hippocampus or amygdala depress the mutual inhibitory system and increase the odour aversion (Bermúdez-Rattoni *et al.* 1983, 1988).

The above evidence suggests that the mechanism introducing odour into the taste system is at least in part regulated by the limbic system. This idea is supported by the finding that temporary or permanent inactivation of the amygdala and hippocampus disrupts taste-potentiated odour aversions, but not formation of simple taste aversion memories. Cholinergic activity of the insular cortex seems to be crucial for the learning of simple conditioned taste aversion. This is in agreement with the finding that CTA training induces a profound increase in tyrosine phosphorylation of several proteins in the insular cortex but not in other brain areas. The presentation of a novel taste and the presentation of the malaise-inducing agent only produced marginal increments of tyrosine phosphorylation. The protein tyrosine phosphorylation has been involved in long-term structural modifications of neurons, underlying learning and memory (Rosenblum *et al.* 1995).

The results discussed above suggest that the limbic system contributes to the association of stimuli indexed with taste and that the insular cortex implements the association of taste with its visceral consequences. Finally, to better understand the mechanism of drug-induced CTA, all the possible physiological actions of any particular drug should be taken into account.

6 Electrophysiology of CTA

Electrophysiological studies provide, together with behavioural effects of lesions and pharmacological experiments, essential information about the neural mechanisms of CTA. This chapter describes the processing of neural information in the projection sites of gustatory (CS) and visceral (US) senses. Although there is no direct electrophysiological evidence for the occurrence of associative learning between gustatory and visceral inputs in single neurons, altered responsiveness to the CS as a result of taste aversion learning has been reported at different levels of the gustatory pathway.

6.1 Taste qualities

It is generally agreed that there are four basic tastes, i.e. sweet, salty, sour, and bitter. In addition to these four tastes, recent studies allow us to include another taste, 'umami', into the basic taste category (Kawamura and Kare 1987; Kawamura *et al.* 1991). Umami is a Japanese word used by Ikeda (1909) to describe the taste of monosodium glutamate (MSG) which he found to be the essence of the savoury taste of sea tangle. An increasing number of taste researchers are now accepting umami as the fifth basic taste.

These basic tastes are also recognized by different species of mammals with a well-known exception that cats do not have taste fibres responsive to sweeteners (Pfaffmann 1941). There are also species differences for the umami taste, i.e. hamsters are almost lacking receptors for umami substances such as MSG and inosine 5-monophosphate (IMP) (Yamamoto *et al.* 1988), rats cannot discriminate umami from sweet tastes (Yamamoto *et al.* 1991), but electrophysiological studies suggest that monkeys can discriminate umami from the other four basic tastes (Baylis and Rolls 1991). Behavioural support for the assertion that mammals can discriminate each of the basic tastes is derived from experiments using the CTA paradigm (e.g. Ninomiya *et al.* 1984; Frank and Nowlis 1989; Yamamoto *et al.* 1991) (Fig. 6.1).

The taste buds, the specialized end organs for gustation, are embedded in the stratified epithelium of the tongue, soft palate, pharynx, larynx, and epiglottis in mammals (Fig. 6.2). Lingual taste buds are associated exclusively with characteristic papillae such as the fungiform, foliate, and circumvallate papillae, while those outside the tongue are found on smooth epithelial surfaces. Each taste bud, which is about 70 μm high and has a diameter of 40 μm, consists of about 100 cells arranged in a compact, barrel-shaped

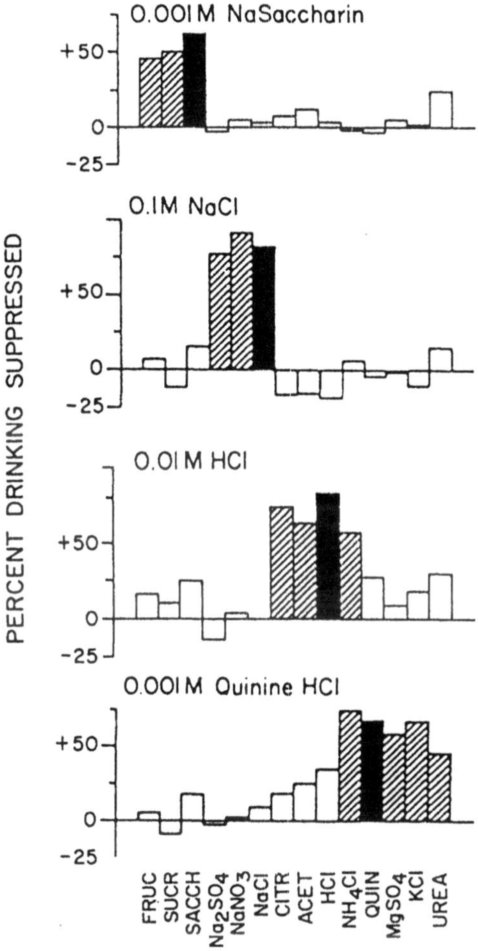

Fig. 6.1 Generalization of conditioned taste aversion among 14 taste stimuli in the hamster. In each graph, a conditioned stimulus is shown on the top, the solid bar represents drinking of the conditioned stimulus, and the striped bars represent significant generalizations of aversion. FRUC, 0.3 M fructose; SUCR, 0.1 M sucrose; SACCH, 0.001 M Na-Saccharin; Na_2SO_4, 0.05 M Na_2SO_4; $NaNO_3$, 0.1 M $NaNO_3$; NaCl, 0.1 M NaCl; CITR, 0.003 M citric acid; ACET, 0.01 M acetic acid; HCl, 0.01 M HCl; NH_4Cl, 0.3 M NH_4Cl; QUIN, 0.001 M quinine HCl; $MgSO_4$, 0.1 M $MgSO_4$; KCl, 0.3 M KCl; UREA, 1.0 M urea. Modified from Frank and Nowlis (1989).

structure (Fig. 6.2). The elongated and bipolar cells in the taste buds are collectively termed taste receptor cells because the apices of these receptor cells are recessed within the taste pore, which is a small opening in the epithelial surface, and tight junctions between the receptor cells restrict the access of the gustatory stimuli to the apical microvilli. A single taste bud is

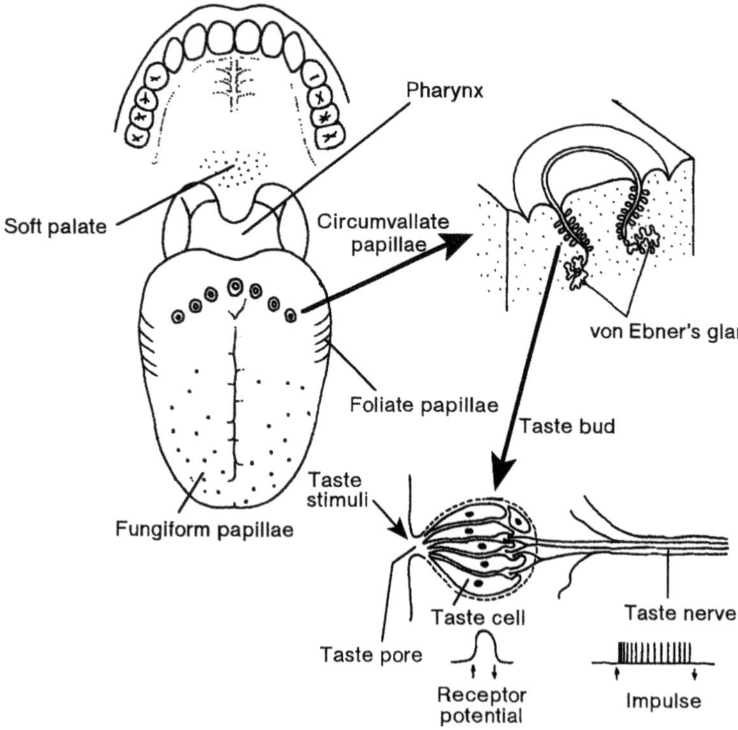

Fig. 6.2 Schematic drawings of locations of the taste buds in the mouth, a cicumvallate papilla, and a taste bud and taste nerves with electrical events.

innervated by more than one nerve fibre and a single nerve fibre can innervate a number of receptor cells situated even in different taste buds or papillae.

The concept of the five basic tastes may be supported by five different transduction mechanisms corresponding to each of the five basic taste stimuli, e.g. sucrose, NaCl, HCl, quinine hydrochloride, and MSG for sweet, salty, sour, bitter, and umami, respectively. Each taste modality has its own specific mechanism (Kurihara *et al.* 1994): salty taste is mediated by Na^+ influx through apical and basolateral Na^+ channels; sour results from H^+ blockade of K^+ channels; sweet is mediated by activation of guanine nucleotide binding protein (G-protein) which is coupled to receptors, leading to adenylyl cyclase-mediated elevation of cAMP; bitter is also transduced via a G-protein-dependent mechanism involving Ca^{2+} release from internal stores; umami may be mediated by metabotropic glutamate receptors.

In laboratory experiments, we use a taste CS with a simple taste without odours to establish CTAs. In the natural environment, however, animals as well as humans encounter various foods, most of which have complex tastes and odours. Therefore, they may use complex mixtures of tastes and/or odours of food components as sensory cues to acquire learned food aversions. This is the reason why they show food-specific aversions rather than taste-specific aversions. Such a conclusion is supported by Bernstein (1978) who reported that cancer patients receiving chemotherapy exhibited aversions to an unusual ice cream eaten just before the drug treatments, but that the aversion did not generalize to other unusual ice creams, which means that the patients did not use sweet taste contained in the ice cream as the sensory cue in the aversion learning, but a complex taste possibly combined with odours.

6.2 Central gustatory pathway

The first step of initiation of taste sensation is stimulation of the taste receptor cells in the taste buds by ions or molecules of food substances and generation of receptor potentials which induce impulse discharges in the four taste nerves (Figs 6.2 and 6.3): the chorda tympani (CT), glossopharyngeal nerve (GN), greater superficial petrosal nerve (GSP), and superior laryngeal nerve (SL). The CT innervates the taste buds in the fungiform papillae on the anterior two-thirds of the tongue; the lingual branch of the GN, those in the foliate and circumvallate papillae on the posterior third of the tongue; the GSP, the palatal taste buds; and the SL, the taste buds on the pharynx, larynx, and epiglottis.

A schematic drawing of the central gustatory pathway in the rat is shown in Fig. 6.3. The taste nerves terminate rostrocaudally in the NTS, i.e. the CT and GSP terminate most densely in the rostral pole, and the GN and SL project more posteriorly from the CT/GSP projection zone extending to the area postrema (Torvik 1956; Beckstead and Norgren 1979; Contreras *et al.* 1982; Hamilton and Norgren 1984). In rats, neurons in the gustatory (rostral) zone of the NTS project to the ipsilateral medial PBN, while those in the viscerosensory (caudal) zone project to the ipsilateral lateral PBN (Norgren 1978; Ricardo and Koh 1978; Herbert *et al.* 1990). The VPMpc of the thalamus, which receives taste inputs from the bilateral PBN, is the synaptic relay of the ascending pathway to the gustatory area of the IC (Yamamoto *et al.* 1980; Yamamoto 1984; Kosar *et al.* 1986; Cechetto and Saper 1987; Ogawa *et al.* 1990). On the other hand, axons from ipsilateral PBN pass ventrally beneath the thalamus giving off fibres to the lateral hypothalamus. They end in the central nucleus of the amygdala and in the bed nucleus of the stria terminalis (Norgren 1976). In monkeys, only the visceral caudal half of the NTS sends axons to the PBN, whereas neurons from the gustatory area of the NTS directly reach the VPMpc and then terminate in the IC (Kinney 1978; Beckstead *et al.* 1980).

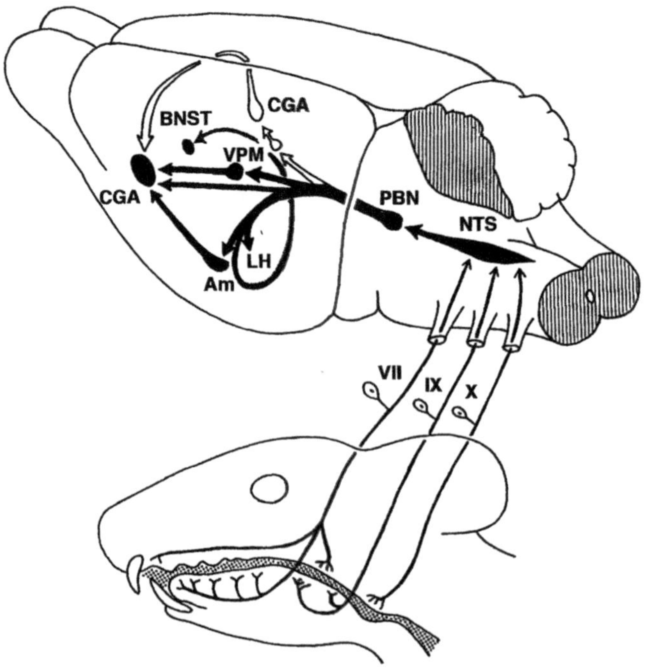

Fig. 6.3 Gustatory afferent pathways in the rat. VII, facial nerve including the chorda tympani and greater superficial petrosal nerve; IX, glossopharyngeal nerve; X, superior laryngeal nerve belonging to the vagal nerve; NTS, nucleus tractus solitarius; PBN, parabrachial nucleus; VPM, ventral posteromedial nucleus of the thalamus; CGA, insular cortical gustatory area; BNST, bed nucleus of the stria terminalis; LH, lateral hypothalamus; Am, amygdala. Ipsilateral projections black, contralateral white. Modified from Yamamoto (1989).

6.3 Processing of taste information

Peripheral taste nerve fibres or central taste-responsive neurons at different levels of the taste pathways do not respond specifically to only one of the basic tastes. Each unit usually responds to more than one of the basic tastes. Such multiple responsiveness to the basic tastes leads to the formulation of the across-neuron pattern (ANP) theory (Erickson 1963). According to this theory, taste quality is coded by the relative amount of activity generated across many neurons.

On the other hand, on the basis of the finding that individual fibres of the hamster and monkey chorda tympani may be categorized into three or four groups that show a best response to one of the four basic taste stimuli, Frank (1973) and Pfaffmann (1974) proposed the labelled-line (LL) theory. According to this theory, information about taste quality is carried by a subgroup of neurons that show greater responsiveness to one of the basic taste qualities than to the others. Recent findings indicate that responses of taste nerve fibres of the chimpanzee tend to respond exclusively to one of the five basic tastes, that gymnemic acid which inhibits sweet taste sensation in humans suppresses responses of only sucrose-best fibres (or fibres which show the most predominant responses to sucrose among the five basic tastes), and that amyloride which inhibits salty taste suppresses responses of only NaCl-best fibres (Hellekant and Ninomiya 1991). These results suggest that each of the basic tastes is conveyed via specific lines of fibres.

Neural substrates for the LL theory may include a topographical arrangement of clusters of neurons with similar response characteristics to taste stimuli. An example of such chemotopical segregation of neurons is suggested in the IC of the rat, i.e. sucrose responses are dominant in the anterodorsal region, NaCl responses in the central and ventral regions, quinine responses in the posterior region, and HCl responses are distributed evenly in the gustatory region of the IC (Yamamoto *et al.* 1985*a*). More recently, Yamamoto *et al.* (1994*a*) have elucidated the functional segregation within the rat PBN by use of c-*fos* immunohistochemistry for anatomical marking of activated neurons: neurons in the external lateral and external medial subnuclei receive taste information for bitter and sour stimuli, neurons in the central lateral subnucleus are related to sweet taste, and neurons in the central medial subnucleus to salty taste (see Chapter 7 for details).

The two theories are not completely different, but each of them seems to stress one aspect of common taste-elicited neural responses, i.e. the importance of a subset of neurons in the processing of taste information (the LL theory) versus the importance of the entire population of neurons (the ANP theory). The above-mentioned effects of gymnemic acid and amyloride can also be explained by the ANP theory, because activation of a specific group of neurons is necessary for the formation of the across-neuron pattern specific to a taste stimulus. Inactivation of a group of neurons also deteriorates an across-neuron pattern, which leads to diminution or alteration of the taste quality. A conceivable interpretation of the two quality coding principles suggests that a gross discrimination of taste categories is quickly achieved by the LL mechanism (Halpern and Tapper 1971) and the subtle difference of taste quality within the same modality is processed in a more time-consuming manner by the ANP mechanism (Scott 1974).

The existence of chemotopic organization in the cortical gustatory area of the rat prompted Yamamoto *et al.* (1985*a,b*) to propose an across-region pattern (ARP) mechanism for taste quality coding at the cortical level, i.e. taste quality is coded by the pattern of the relative amount of neural responses across anatomically discrete areas rather than across individual neurons. The ARP concept explains the qualitative similarity among the taste stimuli found in the behavioural experiment using the CTA paradigm better than the ANP or LL notions (Yamamoto *et al.* 1985*b*). The ARP hypothesis seems to include the key idea of the other two hypotheses. First, the assumption that the chemotopic organization of the cortex represents terminal projection sites of specific channels that convey each of the basic taste qualities leads to the conclusion that the ARP hypothesis accommodates the basic features of the LL hypothesis. Second, the ARP hypothesis may be considered a modified version of the ANP theory since the ARP notion assumes that taste quality is coded by the pattern of the relative amount of neural responses across anatomically discrete areas rather than across the individual neurons as the ANP theory asserts.

6.4 General visceral information

Lithium chloride (LiCl) is one of the most commonly used USs in the CTA paradigm. According to Nachman and Ashe (1973) and Nachman and Hartley (1975), an intraperitoneal (i.p.) injection of isotonic (0.15 M) LiCl (2% of body weight) following a gustatory CS is very effective in establishing strong CTA in the rat. Signs of sickness such as decreased activity, diarrhoea, and urination appear within 5–10 min after injection of LiCl (Nachman 1970; Nachman and Hartley 1975; Di Lorenzo 1985). An electrophysiological study (Yamamoto *et al.* 1989) showed that neurons in the IC of the rat responded tonically to an i.p. injection of 0.15 M LiCl with the shortest latency being approximately 5 min. These findings suggest that LiCl exerts its effect on the peripheral gastrointestinal organs and central nervous system with the latency of 5–10 min.

Niijima and Yamamoto (1994) have recently shown in anaesthetized rats that direct application of 0.15 M LiCl into the intestines elicits gradually increasing and long-lasting discharges in afferent fibres of the vagal (parasympathetic) and splanchnic (sympathetic) nerves with an onset latency of 5–10 min. Their results yielded the following findings. (1) The splanchnic and vagal nerves respond to LiCl administered not only intraperitoneally but also intraduodenally. This finding explains the fact that drinking of LiCl as well as an i.p. injection of LiCl exert US effects (Nachman 1963). (2) The magnitude of neural discharges is larger in the splanchnic nerve than in the vagal nerve both after intraduodenal and i.p. application of LiCl, suggesting a more important role of the splanchnic nerve than of the vagal nerve in transmitting the LiCl-induced sensory information. These results are in agreement with the finding by Martin *et al.* (1978) that bilateral subdiaphragmatic vagotomy in rats did not prevent subsequent acquisition of CTAs when toxicosis was induced with isotonic LiCl. (3) The LiCl activation of C-fibres rather than of B-fibres suggests the nociceptive nature of the information transmitted.

The activation of nociceptive fibres by LiCl application corresponds well with the recent c-*fos* immunohistochemical study showing that after i.p. injections of LiCl (Yamamoto *et al.* 1992*a,b*) or acetic acid (Menetrey *et al.* 1989), c-*fos* immunoreactive neurons were found in the superficial dorsal horn, a projection site for visceronociceptive information in the thoraco-lumbar spinal cord (Sugiura *et al.* 1989). This spinal visceronociceptive information is known to project to the PBN (Hylden *et al.* 1989), a nucleus also receiving gustatory information.

Adachi and Kobayashi (1988) and Adachi *et al.* (1991) demonstrated that some AP neurons responded to direct application of isotonic LiCl. They suggest that excitation of AP neurons by LiCl through a humoral route plays a role in establishing CTA. Ritter *et al.* (1980) and Ladowsky and Ossenkopp (1986) showed that lesions of the AP impaired lithium-induced taste aversion.

Taken together, intraperitoneally injected LiCl exerts its effects as an US with a latency of 5–10 min through the neural routes via the vagus and splanchnic nerves and also through a humoral route.

Recent studies of the visceral sensory system indicate representation of multiple visceral modalities at all levels of the central nervous system (see Cechetto 1987 for review). In the NTS, general visceral afferents are represented caudally, and gustatory afferents rostrally (Adachi 1984, 1994; Chambert *et al.* 1993). At the pontine relay, the PBN, general visceral afferents are represented laterally, but gustatory afferents medially (Cechetto 1987; Yuan and Barber 1991; Suemori *et al.* 1994).

Suemori *et al.* (1994) have shown in the rat that neurons in the lateral PBN respond to visceral stimulation such as distension of the stomach. Yuan and Barber (1991) recorded unit responses in the lateral PBN of the cat to electrical stimulation of gastric vagal and greater splanchnic nerves and suggested that this nucleus receives and processes a substantial amount of general visceral afferent input. The medial PBN (gustatory zone) projects to the parvocellular part of the ventral posterior medial thalamic nucleus, while the more lateral portions of the PBN (general visceral zone) project to the parvocellular part of the ventral posterior lateral nucleus of the thalamus (Cechetto and Saper 1987). These thalamic relays in turn project to the insular cortex. Single-unit recordings in the insular cortex of the rat have demonstrated that neurons in the insular cortex respond to multiple visceral sensory modalities and that there is a viscerotopic organization such that a rostral dysgranular insular zone receives primarily gustatory information whereas general visceral sensation is represented in the caudal granular insular region (Cechetto and Saper 1987; Ito 1992). The hypothalamus and amygdala have also been shown to receive direct viscerosensory projections from the NTS and PBN in the rat. The lateral PBN provides substantial innervation of the central nucleus of the amygdala (Saper and Loewy 1980).

As described above, gustatory information and abdominal visceral information ascend closely together so that it is possible for both inputs to converge on single neurons at all levels of the neuraxis. Recording sites of single unit responses to visceral stimulation seem to overlap with the gustatory responses in the thalamus (Saleh and Cechetto 1993) and in the IC (Ito 1992). Actual convergence of gustatory and abdominal visceral inputs has been reported in the amygdala (Yamamoto and Fujimoto 1991) and IC (Yamamoto *et al.* 1989). Hermann and Rogers (1985) recorded unit responses of the rat PBN to electrical stimulation of the vagal and taste nerves and found convergence of both inputs onto neurons within an interstitial zone in the caudal PBN. The above electrophysiological studies showing projections of gustatory and general visceral signals to the same common levels of the neuraxis together with the recent c-*fos* immunohistochemical studies and behavioural lesion studies have considerably advanced our knowledge of the neural substrate of taste–visceral integration underlying the formation of CTA.

6.5 Alteration of taste responsiveness after CTA

On the basis of electrophysiological characteristics of taste-responsive neurons, it is of great interest to survey changes of taste responsiveness of central neurons to the CS after acquisition of CTA. A specific question is whether CTA learning is accompanied by a change of taste quality of the CS or only by a change of hedonic appreciation of the CS (i.e. the taste quality of the CS is not altered but the CS elicits aversive behavioural responses after CTA).

Since the pioneering studies done by Aleksanyan *et al.* (1976) and Buresova *et al.* (1979), changes of neural responses after the acquisition of CTA have been reported at different levels of neuraxis of the gustatory pathway such as the NTS (Chang and Scott 1984), PBN (Di Lorenzo 1985; Shimura *et al.* 1997*b*), IC (Yamamoto *et al.* 1989), and amygdala (Yamamoto and Fujimoto 1991; Yasoshima *et al.* 1995*a*). The neural mechanisms mediating the response modifications in CTA-trained rats will be a key to understanding the neural basis of CTA. Many of these studies suggest an enhanced response to the CS after the acquisition of CTA, suggesting that the increased responsiveness to the CS is useful for detecting low concentrations of the CS to avoid its ingestion. However, as already suggested by Buresova *et al.* (1979), inhibition of unit activity depending on the location and function of neurons is also frequently induced by presentation of the CS after aversive conditioning.

6.5.1 NTS

Chang and Scott (1984) recorded unit responses of the NTS to taste stimuli in CTA-trained rats and control rats under anaesthetized, paralysed conditions. They found that the saccharin CS evoked a significantly larger response in CTA-trained rats than in control rats, and that the effect was limited to the subgroup of neurons sensitive to sweeteners (Fig. 6.4). These results imply the existence of a functionally distinct channel within the taste system, which is composed of neurons responsive to the sweet quality. Moreover, these alterations of response patterns of the NTS neurons

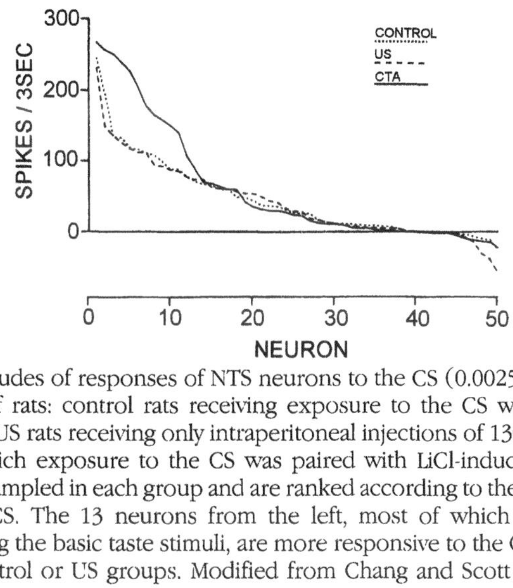

Fig. 6.4 Magnitudes of responses of NTS neurons to the CS (0.0025 M Na-saccharin) in three groups of rats: control rats receiving exposure to the CS without physiological consequences, US rats receiving only intraperitoneal injections of 130 mg/kg of LiCl, and CTA rats in which exposure to the CS was paired with LiCl-inducing poisoning. Fifty neurons were sampled in each group and are ranked according to their responsiveness to the saccharin CS. The 13 neurons from the left, most of which responded best to saccharin among the basic taste stimuli, are more responsive to the CS in the CTA group than in the control or US groups. Modified from Chang and Scott (1984).

probably reflect descending effects from higher centres, because a temporal analysis of responses of the NTS neurons showed that the increased CS response after aversive conditioning started with a fairly long latency of 600–900 ms.

Does the taste quality of saccharin change to a bitter or quinine-like taste after aversion learning to saccharin? This question is raised since saccharin contains a bitter component (Bartoshuk 1979) and c-*fos* immunoreactive labelling induced by saccharin shifts from the dorsal lateral subnucleus, where sweeteners elicit c-*fos* expression, to the external lateral subnucleus, where quinine elicits c-*fos* expression, as the result of learned aversion to saccharin. Chang and Scott (1984) showed that within the three-dimensional taste space, as determined by neural activity in the NTS, the saccharin CS after the acquisition of CTA was in close alignment with quinine. It should be noted here that the taste spaces represent similarities among stimuli in terms of taste hedonics as well as of taste quality. The same paper reported that the CS evoked a significantly larger response in CTA-trained rats than in control rats, and that the effect was limited to the subgroup of neurons sensitive to sweeteners. On the basis of these results, Scott and Giza (1987) have suggested that among the gustatory processes performed in NTS it is primarily the hedonic appreciation of a taste, rather than its quality, that appears to be altered. In fact, behavioural studies have shown that aversion learning to saccharin generalizes only to sweeteners, and not to other taste stimuli representing salty, sour, and bitter tastes, indicating that the taste quality of saccharin does not change from 'sweet' to 'bitter' (Nowlis *et al.* 1980; Frank and Nowlis 1989).

6.5.2 PBN

It is suggested that a subset of PBN neurons shows enhanced responses to a taste stimulus which has been paired with LiCl administration. Di Lorenzo (1985) examined the effects of taste aversion conditioning on unit responses to NaCl in the PBN of rats paralysed with Flaxedil. Initially, units were tested for responsiveness to NaCl, HCl, sucrose, saccharin, and quinine, then units were conditioned by a presentation of NaCl overlapping with an i.v. injection of LiCl. This conditioning procedure resulted in a long-lasting overall increase in the magnitude of the response specifically to NaCl in a subset of PBN units. The conditioned units were more narrowly tuned than units that were not conditioned, but did not necessarily respond best to NaCl. The temporal pattern of response to NaCl did not change after conditioning, i.e. both phasic and tonic portions of the NaCl response increased in magnitude.

Shimura *et al.* (1997*b*) examined the nature of parabrachial responses to 13 taste stimuli in deeply urethane-anaesthetized rats that were pretrained to avoid 0.1 M NaCl in a behavioural paradigm. These responses ($n=53$) were compared to the responses ($n=56$) of naive rats. In both groups more than 80% of units were sensitive primarily to NaCl. Although there were no differences between the mean magnitudes of the responses to each taste stimulus between groups, cluster analysis clearly revealed a strong similarity among responses to sodium salts in the conditioned group (Fig. 6.5). These results suggest that the aversive conditioning modifies PBN units so that salty taste is more salient than other tastes. This would facilitate the gustatory discrimination that is required by conditioned animals.

The existence of centrifugal control of PBN neuronal activity is suggested by the findings that decerebration (Di Lorenzo 1988*b*), infusions of procaine into the IC (Di Lorenzo 1990), or electrical stimulation of the IC (Kiyomitsu *et al.* 1988) all influence responsiveness of PBN neurons. Di Lorenzo's

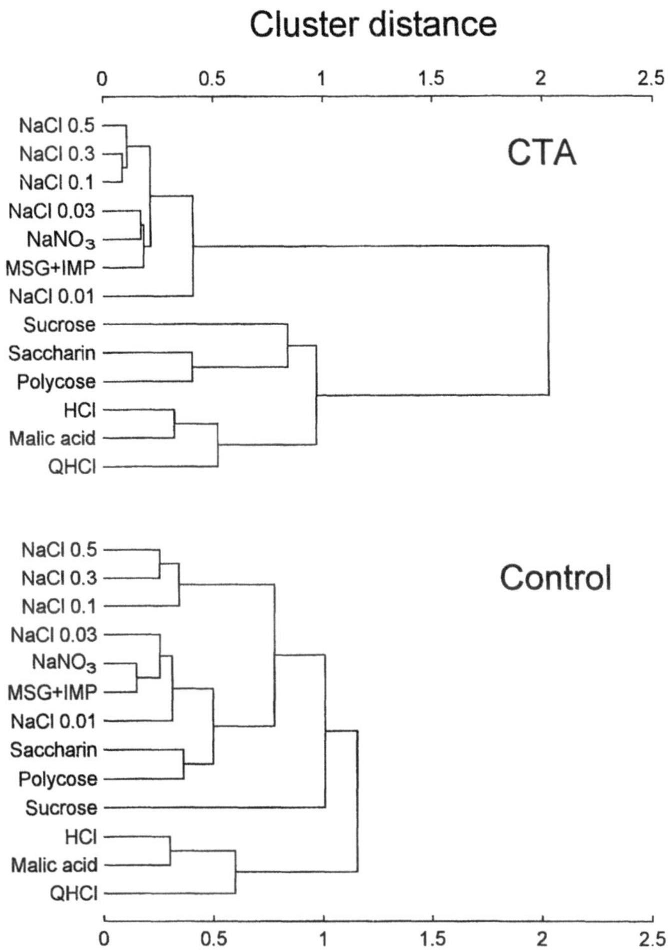

Fig. 6.5 Similarity of taste stimuli as expressed by dendrograms in control rats (control) and in rats trained to avoid 0.1 M NaCl as a CS (CTA). The dendrograms were obtained from a hierarchical cluster analysis of correlation coefficients across unit responses to 13 taste stimuli. Note that similarity among responses to sodium salts are more prominent in the CTA rats than in the control rats. IMP, inosine 5'-monophosphate; MSG, monosodium glutamate; QHCl, quinine hydrochloride. From Shimura *et al.* (1997*b*).

results obtained in paralysed rats may be mediated by influences exerted by the higher centres on PBN activities. However, the data of Shimura *et al.* (1997*b*), obtained in deeply anaesthetized rats, show that the changes of responsiveness which had occurred during the conditioning procedure are retained as a plastic change in the PBN without involvement of higher centres after the acquisition of CTA.

6.5.3 Insular cortex

Based on the responsiveness of neurons in the IC to various taste stimuli in behaving rats, taste-responsive neurons are grouped into a taste-quality type, which reflects the quality coding, and a taste-hedonic type, which differentiates the hedonic (preferable or repulsive) aspect of taste stimuli (Yamamoto *et al.* 1989). Responses of some of the cortical taste-responsive neurons are modified after acquisition of CTA produced by pairing a taste solution with an intraperitoneal injection of LiCl (Yamamoto *et al.* 1989). The effects are different, depending on the type of neuron. The enhanced responses were observed in taste-quality neurons, which are narrowly tuned neurons responsive specifically to a group of tastants eliciting one of the conventional four taste qualities, and a preference response pattern changed to an aversive response pattern after acquisition of CTAs in taste-hedonic neurons, which are defined as neurons showing differential response patterns depending upon hedonics (preference or aversion, acceptance or rejection) of taste stimuli.

More recently, Yasoshima *et al.* (1995*b*) studied the role of the IC in CTA of freely behaving rats by comparing the pre- and post-conditioning unit activities to intraoral infusions of saccharin which was used as a CS. After recording the control responses, rats drank CS for 5 min followed by an i.p. injection of LiCl as a US. At 10 min after the US injection, the rats did not show aversive behaviour to the CS and the CS elicited no significant unit activities. However, 30 min after conditioning, rats showed aversive taste reactivities to the CS and significantly enhanced activities to the CS were observed in 14 (26%) out of 54 units. The enhanced responses in 6 of the 14 units were maintained for more than 6 h. On the other hand, in the other 8 units, the enhanced responses were present for no more than 90 min. These two different time courses of excitatory changes suggest that the IC neurons are involved in formation of both short-term and long-term CTA memories.

6.5.4 Amygdala

The amygdala plays a crucial role in the judgement of hedonic values of sensory stimuli in some instances of emotional learning (LeDoux 1993; Muramoto *et al.* 1993; Helmstetter and Bellgowan 1994). Because the hedonic value of the CS is drastically changed from ingestive to aversive after the acquisition of CTA, the amygdala is assumed to be one of the most plausible sites involved in CTA formation. To investigate the dynamic neural processing within the amygdala in the context of the CTA paradigm, Yamamoto *et al.* (1993*a*) and Yasoshima *et al.* (1995*a*) recorded amygdalar unit activities during voluntary licking or during intraoral infusions of taste solutions, and compared them before and after the conditioning procedure in freely behaving rats (Fig. 6.6).

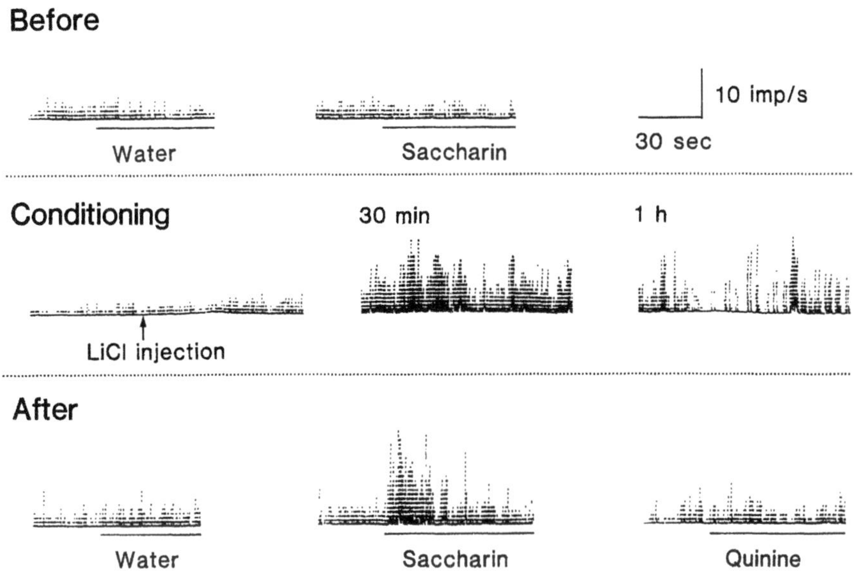

Fig. 6.6 Single unit recording obtained from the basolateral nucleus of the amygdala in response to intraoral infusions of water and taste solutions before, during, and after aversive conditioning to 0.01 M Na-saccharin in the rat. After infusion of saccharin, 0.15 M LiCl (2% of body weight) was i.p. injected for establishment of CTA. From Yamamoto *et al.* (1993).

According to their results, only a limited number of amygdalar units (9 of 137 units)were responsive to taste stimuli before conditioning. The US (an i.p. injection of 0.15 M LiCl) elicited excitatory and/or inhibitory responses in another 9 of 137 units. These responses lasted less than 2 h. These LiCl-responsive units did not respond to any taste stimuli before and after conditioning. When tested 3–4 h after pairing saccharin (CS) with the US, 32 of 137 units responded to the CS: 20 of these 32 units were facilitatory-type and 12 were inhibitory-type (Fig. 6.7).

The facilitatory-type units, which responded to the CS in an excitatory direction after CTA, were found more frequently in the BLA ($n=17$) than in the CeA ($n=3$). The facilitatory-type units showed a response pattern similar to that of the taste-quality type. It is suggested that the BLA plays important roles in aversive or avoidance behaviour, such as neophobia (Nachman and Ashe 1974), fear conditioning (LeDoux 1993; Helmstetter and Bellgowan 1994), passive (Yajeya *et al.* 1991) and discriminative (Maren *et al.* 1991; Muramoto *et al.* 1993) avoidance learning, and CTA (Yamamoto *et al.* 1995*a*). The facilitatory responses in the BLA, therefore, may be related to the conditioned aversive responses, but not to such innate aversive reflexes as elicited by unpalatable stimuli, HCl or QHCl, because the latter stimuli did not induce any remarkable responses in these units.

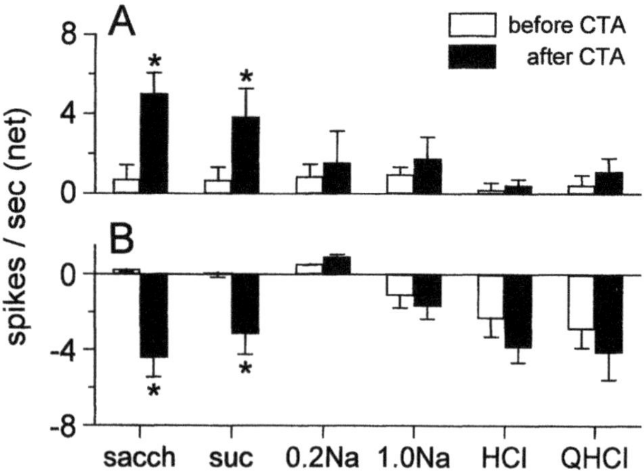

Fig. 6.7 Two types of amygdalar units exhibiting facilitatory or inhibitory responses to 0.005 M Na-saccharin CS after CTA. (A) Mean responses of 20 facilitatory-type units. The responses to saccharin (sacch) and 0.5 M sucrose (suc) were enhanced significantly after conditioning (solid bar) in comparison to those before conditioning (open bar). No significant differences were detected in the responses to 0.2 M NaCl (0.2 Na), 1.0 M NaCl (1.0 Na), 0.03 M HCl (HCl), and 0.001 M quinine hydrochloride (QHCl). (B) Mean responses of 12 inhibitory-type units. The responses to the saccharin CS and sucrose were inhibited significantly after conditioning in comparison to those before conditioning. From Yasoshima *et al.* (1995*a*).

On the other hand, the inhibitory-type units, which responded to the CS in an inhibitory direction after CTA, were found more frequently in the CeA ($n=11$) than in the BLA ($n=1$). Aversive taste stimuli such as HCl and QHCl also inhibited the spontaneous activity of these units regardless of the CTA procedure, indicating that these units belong to the taste-hedonic type. The occurrence of such inhibitory responses to the hedonically negative tastes suggests that the normal function of the CeA is to stimulate ingestive behaviour. This suggestion is supported by the recent finding (Minano *et al.* 1992) that inhibition of CeA neural activities by microinjection of muscimol, a GABA -selective receptor agonist, into the CeA produced a decrease of food intake. Alternatively, these inhibitory units might be related to the taste-elicited autonomic responses because it is known that the CeA is involved in the autonomic conditioned responses such as heart rate conditioning (Kapp *et al.* 1992) and conditioned cephalic insulin response (Roozendaal *et al.* 1990).

The above-mentioned electrophysiological findings suggest that recognition of the CS and expression of rejection behaviour after the acquisition of CTA is based on an interplay between the BLA and CeA, i.e. qualitative and cognitive aspects of the CS are processed in the BLA and hedonic aspects in the CeA. Behavioural lesion studies have shown that lesions of the BLA severely disrupt the acquisition and retention of CTA, while lesions of the CeA have no effect on the CTA formation (Yamamoto *et al.* 1995*a*). Since the BLA has dense excitatory fibre connections to the ventromedial nucleus of the hypothalamus (VMH) (Oomura and Ono 1982), which is known to be the 'satiety centre', excitation of VMH from the BLA may play a role in avoidance behaviour. A more important role of the BLA is to evaluate the biological significance of sensory stimuli for the animals (Ono and Nishijo 1992). Without the BLA, therefore, CTAs are disrupted because the animals cannot associate the CS with the US during acquisition or recognize the aversive nature of the CS during retrieval.

6.5.5 Hypothalamus

Hypothalamic neurons also change their response patterns in terms of the hedonic dimension, i.e. saccharin (CS) activates the feeding centre and inhibits the satiety centre in naive animals, but elicits opposite effects after CTA acquisition to the CS (Aleksanyan *et al.* 1976). Since the hypothalamus receives inputs from the IC and the amygdala, altered responsiveness of hypothalamic neurons may simply reflect that of IC and/or amygdala neurons which as described above exhibit altered responses to the CS after CTA acquisition.

7 Functional morphology of CTA

Our understanding of the neural mechanisms of CTA can be essentially advanced by identification of the cellular and subcellular events occurring during association, acquisition, consolidation, retention, and retrieval phases of CTA memory. This requires, on the one hand, simultaneous recording of the activity of many neurons throughout the whole brain, and on the other hand, detection of subcellular, molecular, or genetic events taking place during each phase of CTA. Although such effort may still be premature, several attempts have been made to use for these purposes immunohistochemical detection of expression of a family of proto-oncogenes and pharmacological manipulation of intracellular second messenger pathways.

To elucidate the functional neuroanatomy of circuits implementing the formation of CTA, it is helpful to use an anatomical marker for activated neurons. The early expression of c-*fos* and other immediate early genes within neurons has been exploited as a means for mapping the neurons responding to sensory stimuli. One of the most useful methods for such a purpose is immunohistochemical staining of the Fos protein and Fos-related antigens, products of c-*fos* expression. c-*Fos* is a proto-oncogene which is expressed in cells within a few hours following voltage-gated calcium entry into the cell after the application of a stimulus (Morgan and Curran 1986). Fos and Fos-related proteins are expressed for more prolonged periods of time, are readily revealed by immunohistochemistry, and can be utilized to identify and localize neuronal activity (Fig. 7.1). Recently, several researchers have been trying to identify in the brain neurons showing c-*fos*-like immunoreactivity (c-*fos* neurons) after visceral stimulation with an i.p. injection of LiCl and after oral stimulation with taste solutions of different qualities and different hedonic values in animals with or without CTA. Fos immunostaining may prove useful in defining brain pathways involved in acquisition and expression of CTA.

Possible uses of other anatomical markers and intracellular events occurring in relation to CTA are also discussed.

7.1 c-*fos* expression after LiCl injection

The distribution of c-*fos* neurons has been examined in the rat brain after an i.p. injection of various concentrations of LiCl including isotonic (0.15 M) LiCl which is commonly used to induce malaise in the CTA paradigm (Yamamoto *et al.* 1992*a,b*; Gu *et al.* 1993). c-*fos* neurons, which are induced

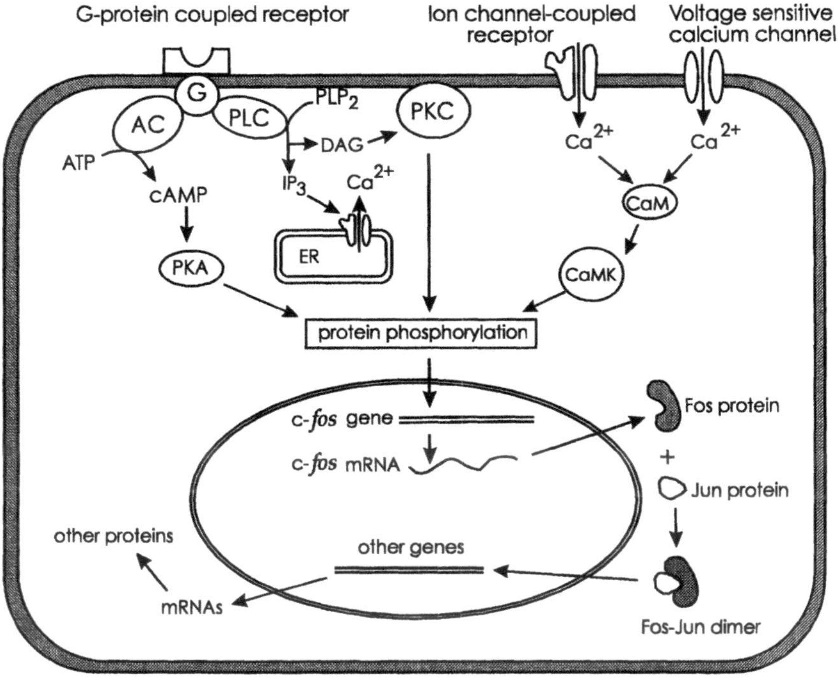

Fig. 7.1 Schematic drawing showing transcription mechanism of a proto-oncogene, c-*fos*, through second messengers triggered by stimulation to a cell.

most dominantly 2–4 h after i.p. injection of LiCl, have been found in the superficial layer of the dorsal horn of the thoracolumbar spinal cord, lateral reticular nucleus, AP, commissure and intermediate parts of the NTS, PBN, locus coeruleus, inferior colliculus, central gray, dorsal raphe, amygdala, supramammilary nucleus, thalamic paraventricular nucleus, hypothalamus (posterior, lateral, and preoptic areas and dorsomedial, paraventricular, and supraoptic nuclei), piriform cortex, septal nucleus, insular cortex, and neocortex. When the amount of LiCl injected is varied, a dose-dependent increase of c-*fos* expression was demonstrated in the AP, NTS, amygdala, and hypothalamic paraventricular and supraoptic nuclei. c-*fos* expression in some areas among the above-mentioned structures such as the paraventricular nucleus of the thalamus, inferior colliculus, locus coeruleus, piriform cortex, and neocortex may be interpreted as the result of general factors irrelevant to the effect of LiCl, e.g. stress induced by handling and injections could be one of these factors.

There are some differences in c-*fos* distribution reported: Yamamoto *et al.* (1992*a,b*) found strong c-*fos* expression in the AP, especially in its rostral portion, Swank *et al.* (1995) showed a moderate amount of c-*fos* expression, but Gu *et al.* (1993) found very few, if any, c-*fos* neurons in the AP, and

Swank and Bernstein (1994) did not observe any c-*fos* neurons in the AP after i.p. injection of LiCl. Moreover, Yamamoto *et al.* (1992*a,b*) observed many c-*fos* neurons in the central amygdaloid nucleus, but Gu *et al.* (1993) observed them in the basolateral amygdaloid nucleus. One major reason for these inconsistencies in c-*fos* immunoreactivity may be accounted for by variable specificity of the commercially available antibody.

Afferent signals elicited by an i.p. injection of LiCl have been shown to be transmitted through humoral and neural routes. The ascending route transmitting LiCl-induced information may involve the sympathetic splanchnic nerve as well as the parasympathetic vagal nerves. Yamamoto *et al.*. (1992*b*) compared LiCl-induced c-*fos* expression among naive, vagotomized, AP-lesioned, i.v. administered, and i.p. administered rats, and suggested that NTS neurons are activated via bilateral vagus nerves, while AP neurons, humorally as well as neurally, via the vagal nerve with right-side predominance. The activated NTS and AP neurons project mainly to the external lateral subnucleus of the PBN and less to the central subnucleus of the PBN. As far as the humoral transmission of LiCl is concerned, Sandner *et al.* (1994) found that the circumventricular structures such as the AP, periventricular, periaqueductal gray and lateral septum contained the highest concentration of Li^+ when the distribution of lithium was directly mapped in the rat brain after an injection of LiCl.

c-*Fos* is less expressed in the thoracolumbar spinal cord than in the NTS (Yamamoto *et al.*. 1992*b*), although strong discharges are observed in the sympathetic splanchnic nerve as well as in the parasympathetic vagus nerve (Niijima and Yamamoto 1994). Traub *et al.* (1994) reported a similar finding: gastric distension and oesophageal distension induced considerable c-*fos* expression in the NTS and virtually none in the thoracic spinal segments. They suggest that parasympathetic visceral afferents may differ from sympathetic visceral afferents by their ability to induce c-*fos* following noxious visceral distension.

More recently, c-*fos* expression in rat brain was examined by an i.p. injection of D-amphetamine sulphate (Swank *et al.* 1995). This drug is also an effective US in the formation of CTA. However, the mechanism of action of this drug and LiCl is quite different in terms of the AP and related emetic circuitry which are critical to the response to LiCl but not to amphetamine. According to Swank *et al.* (1995), the most dramatic elevations in c-*fos* immunoreactivity were seen in intermediate NTS and slight elevations in the AP and caudal NTS by administration of these drugs. In these three regions, the response to LiCl was more than twofold to threefold greater than that seen after amphetamine (Fig. 7.2A). In the lateral PBN amphetamine and LiCl induced comparable and significant increases in c-*fos* expression (Fig. 7.2B). Amphetamine also induced many c-*fos* neurons in striatum, while no c-*fos* neurons were visible here after LiCl. If other USs with different characteristics such as non-emetic drugs, irradiation, or rotation of the body should induce c-*fos* expression in a certain common area of the brain, there is a high probability that this is the critical site for the formation of CTAs. Sakai and Yamamoto (1997) suggest that such a region is the lateral portion of the PBN.

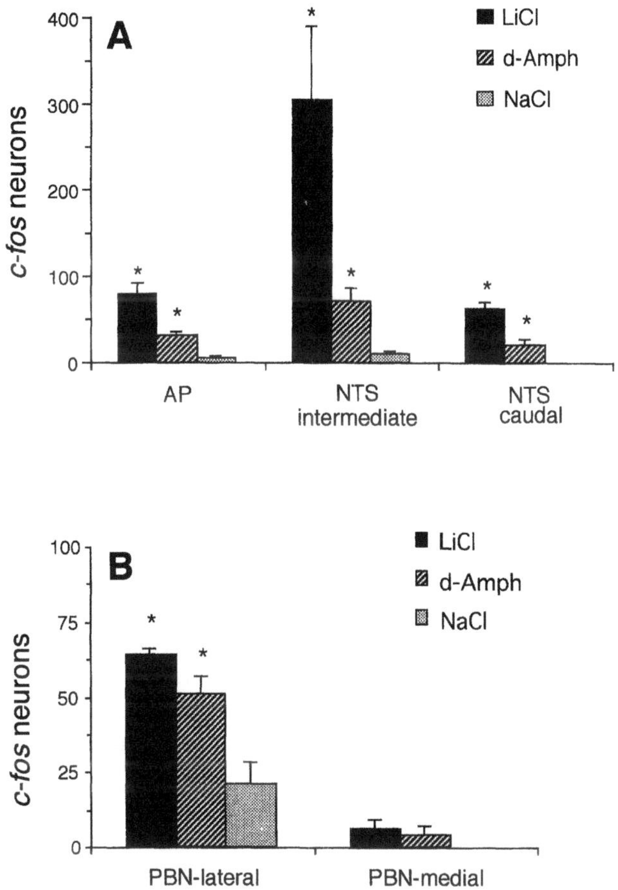

Fig. 7.2 Mean number of c-*fos* neurons in the area postrema (AP) and nucleus of tractus solitarius (NTS, intermediate and caudal divisions) (A) and pontine parabrachial nucleus (PBN, medial and lateral divisions) (B) in rats. Intraperitoneal injections of either LiCl (0.15 M, 20 ml/kg) and D-amphetamine sulphate (3 mg/kg, d-Amph) induced elevation of c-*fos* expression relative to saline (20 ml/kg, NaCl) injected controls. The most marked elevations were seen in the intermediate NTS for LiCl-injected rats. Both LiCl and amphetamine induced comparable and significant increases in c-*fos* neurons relative to saline controls. *Significant at $P < 0.05$ relative to NaCl. Modified from Swank *et al.* (1995).

7.2 c-*fos* expression after taste stimulation

The c-*fos* immunohistochemical method could be a useful tool for localization of neurons in the brain responding specifically or non-specifically to a variety of taste stimuli with different qualities and hedonic values. This possibility was explored by Harrison and Miller (1994) by showing that electrical stimulation of the two taste nerves, chorda tympani and glossopharyngeal nerves, elicited c-*fos* expression in the NTS and PBN, but not in the thalamus. Travers *et al.* (1994) and Harrer and Travers (1996) found c-*fos* neurons in the NTS in rats responding to 1 M sucrose and/or 0.003 M quinine, and have detected a rough chemotopic localization of c-*fos* expression between these two stimuli within the NTS and have suggested that the topographic pattern of taste afferent termination in the NTS is related to their chemosensitivity as well as their peripheral spatial distribution, confirming their electrophysiological study disclosing a clear orotopic organization within this nucleus. c-*fos* immunohistochemistry may not be a sufficiently sensitive anatomical marker for NTS taste-responsive neurons. Yamamoto *et al.* (1993*b*, 1994*a*), on the other hand, succeeded in detecting c-*fos* expression elicited by taste stimuli within the PBN of the rat. Taste-elicited c-*fos* expression in other parts of the brain has not yet been well analysed.

The distribution of evoked expression of c-*fos* was immunohistochemically examined in the PBN of water-deprived rats after free ingestion of palatable liquids, after intra-oral infusion of aversive taste solutions, and after an i.p. injection of LiCl (Fig. 7.3) (Yamamoto *et al.* 1993*b*, 1994*a*). Abdominal visceral information induced by an i.p. injection of LiCl projects to the rostral part of the external lateral subnucleus (els). Caudally, taste information corresponding to quinine and HCl projects to the lateral border of the brachium in a 'capping' pattern including the els and external medial subnucleus (ems). More caudally, hedonics of taste may be represented, i.e. the dorsal lateral subnucleus (dls) may be related to ingestive behaviour or positive hedonics, and the els may be related to aversive feeling or negative hedonics. Most caudally, taste information elicited by sweeteners and NaCl may project to the central lateral subnucleus (cls) and central medial subnucleus (cms), respectively. Generally speaking, the medial part of the PBN, which consists of dls, cls, and cms, seems to be concerned with palatable tastes and hedonically positive value, while the lateral part, which consists of els and ems, is related to aversive tastes with hedonically negative value and gastrointestinal distress.

7.3 c-*fos* expression in the PBN after CTA

As described above, saccharin induced c-*fos* neurons in both dls and cls in naive rats. On the other hand, rats with CTAs to saccharin showed c-*fos* neurons predominantly in the els as well as in the cls of the PBN after intra-oral infusion of saccharin (Fig. 7.4) (Yamamoto 1993*b*, 1994*a*). These results suggest that the recipient zone for saccharin taste shifts from the dls to els after acquisition of CTAs, although saccharin-induced c-*fos* neurons in the cls appear in the aversively conditioned rats as well as in naive rats. In other words, the hedonic aspect of saccharin taste shifts from positive to negative while the quality aspect remains unchanged after aversive conditioning to saccharin. Essentially the same phenomenon can be seen when NaCl is used instead of saccharin (Yamamoto *et al.* 1994*a*).

It is possible that activation of neurons in the dls and els of the PBN is due to

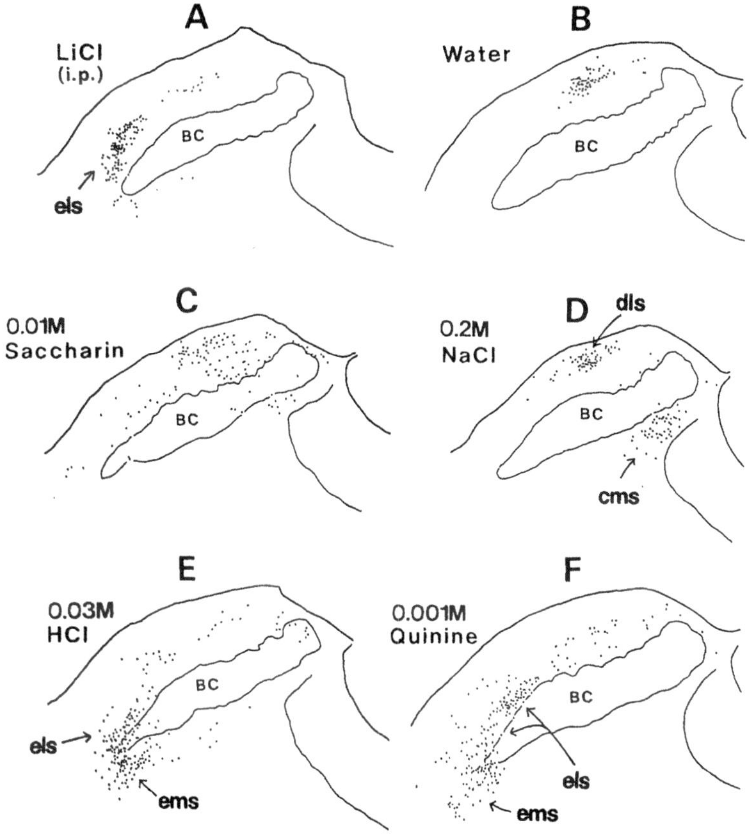

Fig. 7.3 Camera-lucida drawings of c-*fos* neurons (dots) in the parabrachial nucleus taken from a representative section after ingestion of distilled water, Na-saccharin, or NaCl, after an intraoral infusion of HCl or quinine hydrochloride, and after an intraperitoneal (i.p.) injection of LiCl in the rat. els, external lateral subnucleus; dls, dorsal lateral subnucleus; cms, central medial subnucleus; ems, external medial subnucleus, BC, brachium conjunctivum. Taken from Yamamoto *et al.* (1994*a*).

descending influences from higher brain centres rather than to exclusive ascending inputs. Namely, the shift of saccharin projections from the dls to the els after CTA acquisition may be mediated by higher brain sites. To elucidate this point, a c-*fos* immunohistochemical study (Yamamoto 1994*a*) was performed in rats with combined large ibotenic acid lesions of the cortical taste area, thalamic taste area, and/or amygdala. Since the rats with the combined lesions cannot acquire CTAs (Yamamoto *et al.* 1995*a*), they drank the saccharin solution even after its intake had been paired with an i.p. injection of LiCl. When such animals ingested saccharin, c-*fos* neurons were weakly observed in the els corresponding to the hedonically negative value, but not in the dls, of the PBN. These results suggest that (1) the hedonic aspect of gustatory information can be processed, even if not completely, at the PBN level without involvement of the higher levels of the central nervous system, and (2) even if the hedonic shift has occurred at the PBN level after the CTA procedure, expression of aversive and avoidance behaviour requires the higher brain centres. The latter statement is consonant with the finding by Grill and Norgren (1978*c*) that chronically decerebrate rats neither rejected nor decreased ingestive reactions to a novel taste after that taste had been repeatedly paired with LiCl-injected illness.

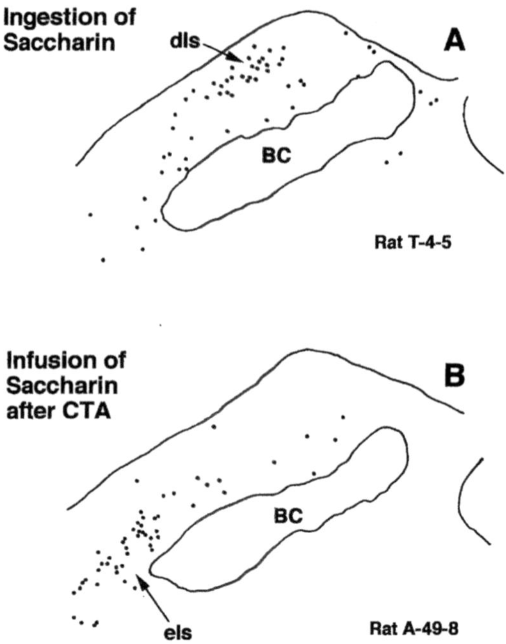

Fig. 7.4 Camera-lucida drawings of c-*fos* neurons (dots) in the parabrachial nucleus of the rat after free ingestion of 0.01 M Na-saccharin (A) and an intraoral infusion of 0.01 M Na-saccharin after acquisition of CTA (B). Note the shift of saccharin-induced c-*fos* expression from the dorsal lateral subnucleus (dls) to the external lateral subnucleus (els) after aversive conditioning.

In decerebrate rats, which received a bilateral knife cut at the precollicular level, c-*fos* expression was similarly observed in the NTS and PBN in response to gustatory stimuli such as sucrose, saccharin, and quinine solutions and to an i.p. injection of LiCl as in intact rats. However, the shift of c-*fos* expression from the dls to els after CTA acquisition to sucrose and saccharin was not so marked in decerebrate rats as was seen in intact rats (Yamamoto *et al.* 1995*b*). These results again show that qualitative and hedonic aspects of gustatory information can be processed within the lower brainstem, but that adequate hedonic change after CS–US association requires descending influences from the higher levels of the central nervous system.

7.4 CS–US association in the PBN

On the basis of differences in distribution of c-*fos* neurons in naive and CTA-trained rats, a hypothesis has been proposed for the integration of gustatory and visceral signals within the PBN (Fig. 7.5) (Yamamoto 1993). When the rat ingests saccharin as the CS, saccharin-derived neural information projects primarily to the cls for taste quality processing, to the dls for positive hedonics, and weakly to the caudal els for negative hedonics. Gustatory signals of any taste stimulus project to the hedonically positive site, negative site, and quality transmission site with different projection densities depending on the particular taste stimulus. When the rat receives an LiCl injection as the US during ongoing activity of the CS, neural information of injected LiCl projects primarily to the rostral els for transmission of visceral sensation, to the caudal els for negative hedonics, and weakly to the dls for positive hedonics. The US and CS inputs converge on neurons in the els and the US signal potentiates the CS transmission in a long-term manner, whereas neurons receiving the convergent inputs in the dls show depressed CS transmission by the US signal in a long-term manner.

The possible existence of such neurons co-activated by both taste and general visceral stimuli has been suggested by Hermann and Rogers (1985) in the interstitial zone of the posterior PBN, but neither in the els nor dls. More extensive studies are needed to explore the existence of co-activated neurons to both taste and visceral stimulation and their location within the PBN. After acquisition of CTAs, therefore, even if the CS information projects similarly, the neural activity in the hedonically negative site (els) increases and the activity in the hedonically positive site (dls) decreases, whereas the activity of quality transmission neurons in the cls is unchanged. Such changes in neural responses in the dls and els may be represented as a shift of c-*fos* neurons from the dls to els as the result of formation of CTAs.

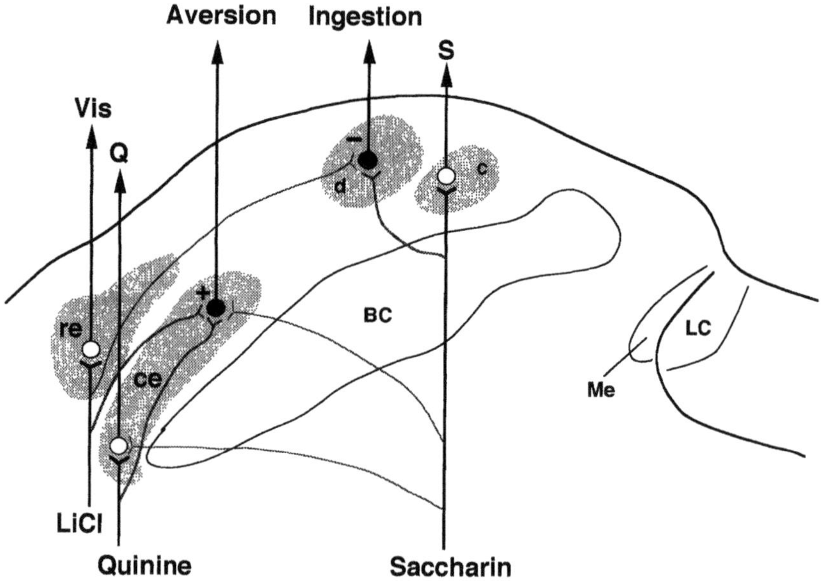

Fig. 7.5 Diagram depicting possible interactions of general visceral and taste information in the parabrachial nucleus. Sensory information elicited by i.p. injection of LiCl projects mainly to quality neurons (open circle) in the rostral part of the external lateral (re) subnucleus and to hedonic neurons (solid circle) in the caudal part of the external lateral (ce) and dorsal lateral (d) subnuclei. Taste information of saccharin projects mainly to quality neurons in the central lateral (c) subnucleus and to other quality neurons and hedonic neurons in the d and ce. BC, brachium conjunctivum; Me, trigeminal mesencephalic nucleus; LC, locus coeruleus; Vis, visceral sense; Q, quinine taste; S, saccharin taste; +, long-term potentiation; -, long-term depression. Modified from Yamamoto (1993).

It is conceivable that saccharin information projects to the els because saccharin contains a 'bitter', i.e. slightly aversive, component (Bartoshuk 1979) and a learned quinine aversion generalizes to saccharin (Morrison and Jessup 1977). This model, therefore, is applicable to saccharin but not to sucrose which is also sweet but is supposed to have only a small aversive component. However, a similar shifting of c-*fos* expression elicited by sucrose stimulation is observed after the rat has learned aversion to sucrose. It is assumed that every taste projects to the els and dls in addition to its quality projection site and that the model is applicable to every taste. The finding (Braun *et al.* 1972) that quinine is more easily associated with illness than saccharin may reflect the more abundant projections to the els for quinine than for saccharin.

The interaction of gustatory and general visceral signals has been addressed by Blake and Lin (1978) and Cabanac (1979) who have indicated that visceral information has the ability to bias or 'switch' the affective aspect of a gustatory stimulus, i.e. pleasure or displeasure, and the consequent ingestive behaviour, i.e. acceptance or rejection. One can assume that such a gustatory–visceral association primarily takes place at the lower brainstem level considering that CTAs (more precisely, the association of gustatory trace and malaise) can be acquired under deep anaesthesia (Buresova and Bures 1977; Rabin and Rabin 1984; Bures and Buresova 1989), under deep hypothermia (Ionescu and Buresova 1977*b*) assumed to elicit functional decerebration, or under cortical spreading depression (Buresova and Bures 1973). It should be kept in mind, however, that no CTAs can be acquired when the CS is applied under these conditions as shown by Bures and co-workers (Buresova and Bures 1973; Buresova and Bures 1977; Bures and Buresova 1989; Ivanova and Bures 1990*a,b*), indicating the importance of descending influences from the higher centres in priming the brainstem association mechanism.

Thus, the neural substrate for associative taste aversion learning can be implemented by the convergence of CS and US inputs in the PBN. Important in this context is the observation by García and Koelling (1966) who have shown that taste stimuli can be easily associated with the internal consequences of X-ray irradiation or LiCl injections, but not with noxious electrical stimulation of the paws. After repeated pairings of taste with electrical shock, however, rats can discriminate a taste stimulus from water, tasty vs. tasteless (Spector *et al.* 1990). They can avoid consuming sucrose without showing aversive reactions when tested with the taste reactivity test (Pelchat *et al.* 1983). Nachman demonstrated that orosomatic stimuli such as a change in the temperature of distilled water (Nachman 1970) or a change in the tactile stimulation of the tongue corresponding to lapping vs. licking or to licking from a spout with a large vs. small opening (Ashe and Nachman 1980) could serve as sufficient CSs when followed by LiCl injections. These results are in agreement with the neuroanatomical and electrophysiological findings indicating that orosomatic inputs transmitted through the trigeminal nerve project to the PBN (Ogawa *et al.* 1982; Cechetto *et al.* 1985). On the other hand, it is assumed that PBN neurons receiving convergent inputs from the gustatory and somatosensory afferents from the body surface are few in number.

Since the PBN and NTS are centres of gustatory reflex responses, some behavioural features of CTAs may be implemented by the local PBN circuitry, i.e. although CTA formation requires participation of post-PBN structures (Grill and Norgren 1978*c*), some behavioural manifestations of CTAs may be a result of a plastic change in the PBN. For example, Halpern and Tapper (1971) showed that rats conditioned to avoid a CS (0.3 M NaCl) could discriminate the CS within a brief period (250 to 600 ms) after stimulus onset. Pre-absorptive insulin secretion, which is known to be reflexively elicited by intraoral infusions of glucose in chronic decerebrate rats (Grill and Berridge 1981), disappears after aversive conditioning to

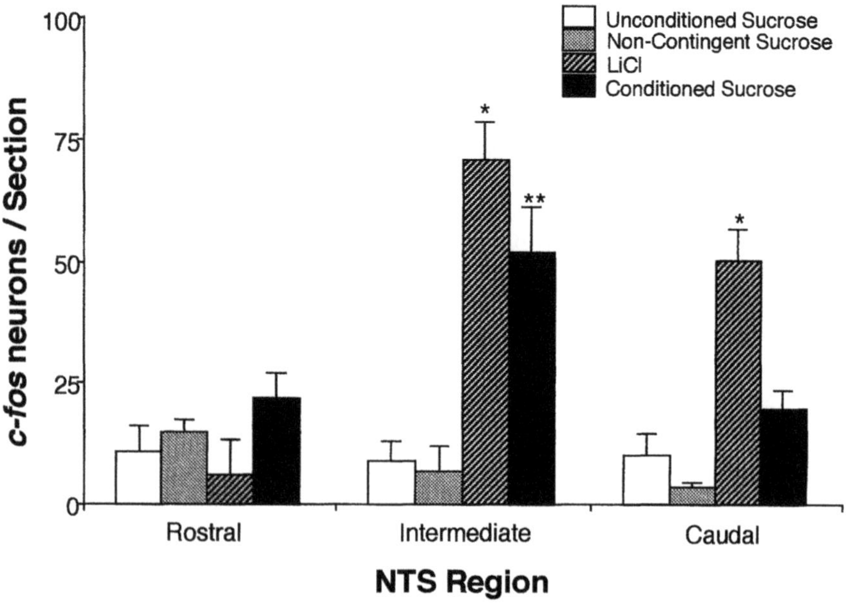

Fig. 7.6 Mean number of c-*fos* neurons in rostral, intermediate, and caudal subregions of the nucleus tractus solitarius (NTS) in four groups of rats. LiCl injections induced significantly more c-*fos* expression in intermediate and caudal subregions than sucrose infusions to unconditioned rats. Sucrose infusions in conditioned rats induced significantly more c-*fos* expression only in intermediate subregion than those in unconditioned rats. Significant at *$P < 0.05$ and **$P < 0.005$ relative to unconditioned sucrose. Modified from Houpt *et al.* (1994).

glucose (Berridge *et al.* 1981). Such a quick reaction time and such dramatic changes of the CS-elicited reflexes after CTA formation may result from switching the activated neurons from positive hedonics to negative hedonics at the PBN level.

7.5 c-*fos* expression in the NTS after CTA

Plastic changes similar to those as described above may or may not occur in the NTS since the projection zones for taste and gastrointestinal inputs are separated rostrocaudally (Norgren 1978; Hermann *et al.* 1983; Cechetto 1987).

The pattern of neuronal activation in the NTS in response to a gustatory CS was examined using c-*fos* immunoreactivity before and after CTA formation in the rat. Houpt *et al.* (1994) examined the responsiveness of the NTS to 6 ml of 5% sucrose delivered by intraoral catheter before and after pairing with systemic LiCl injections. They found that sucrose infusion alone did not induce c-*fos* neurons anywhere in the NTS or AP, LiCl alone induced strong c-*fos* neurons in the caudal and intermediate portions of the NTS as well as in the AP, and oral infusion of sucrose after CTA formation induced c-*fos* neurons in the intermediate portion of the NTS (Fig. 7.6). Since no c-*fos* neurons were induced in the rostral (gustatory) portion by sucrose infusion after CTA and in the intermediate portion by aversive quinine infusion, they suggest that c-*fos* expression observed in the intermediate portion reflects NTS activation by conditioned visceral, cardiac, respiratory, or stress responses to sucrose infusion which are correlated with a subset of responses produced by LiCl. Houpt *et al.* (1995) believe that the c-*fos* induction correlated with CTA expression was not dependent on subdiaphragmatic vagal afferent input, because there was no significant difference in the number of c-*fos* neurons in the intermediate NTS between vagotomized and sham-vagotomized rats re-exposed to sucrose CS.

Essentially similar results were reported by Swank and Bernstein (1994) and Thiele *et al.* (1996). Their major finding was that re-exposure to saccharin induced c-*fos* in the intermediate portion of the NTS which is also activated by the LiCl or ethanol in rats that had received a single pairing of the saccharin CS with the LiCl or ethanol US, but not in unconditioned controls. These results may support the behavioural finding that, after taste aversion conditioning, re-exposure to the taste CS can promote symptoms of conditioned illness such as the lying on belly posture (Meachum and Bernstein 1990).

The location in the intermediate NTS, where c-*fos* expression was observed in CTA-trained rats after re-exposure to the CS, is well suited for integration of the information from the rostral gustatory and caudal visceral portions of the NTS. Thus, conditioned gastrointestinal responses forming a subset of the unconditioned effects of LiCl, e.g. diarrhoea, lying prone, and piloerection, are induced at the NTS level. The hedonic shift described above occurs at the PBN level, and the association between taste quality and hedonic value may be established at the more rostral level, probably in the amygdala (Yamamoto 1994).

More recent studies (Houpt *et al.* 1995; Swank *et al.* 1995; Houpt *et al.* 1996) suggest, however, that c-*fos* expression in the intermediate NTS is a correlate of cellular events involved in learning, rather than a pattern of neural activation mimicking the effect of the US. This notion is supported by the following arguments. (1) LiCl induced dense c-*fos* immunoreactivity in the intermediate NTS while amphetamine induced only a light c-*fos* expression in this area. The c-*fos* expression in NTS after saccharin–amphetamine pairing elicited a pattern quite similar to that seen after saccharin–LiCl pairing (Swank *et al.* 1995). (2) The saccharin CS induced significant elevations of c-*fos* expression in NTS of CTA-trained rats whereas quinine did not, although behaviourally the animals' ingestive responses were quite similar. This suggests that the c-*fos* response in NTS is not a non-specific occurrence of the aversiveness or stressfulness of this experience (Swank *et al.* 1995). (3) The failure of vagotomy to affect c-*fos* induction in NTS after CS re-exposure rules out the possibility that the c-*fos* induction is activated by the vagal afferent input (Houpt *et al.* 1995). (4) In view of the essential role of the PBN, amygdala, and insular cortex in CTA, it is conceivable that conditioned c-*fos* induction reflects descending influences which may prime processing of the gustatory CS in the NTS relay neurons (Swank *et al.* 1995).

In consonance with this suggestion, Schafe *et al.* (1995) used chronic hemidecerebrate rats, which have a unilateral brain transection at the level of the superior colliculus, and found that the unconditioned c-*fos* immunoreactivity to LiCl was unaltered by the transection, while the conditioned expression of c-*fos* immunoreactivity to the CS taste was evident only on the side of the NTS which retained neural connections with the forebrain, implicating forebrain input in this cellular correlate of CTA learning. (5) Intraoral infusions of quinine induced c-*fos* immunoreactivity in the intermediate NTS, which was quantitatively and anatomically similar to that elicited by the sucrose CS re-exposure, only after acquisition of CTA against quinine, but not in unconditioned, non-contingently treated, or extinguished rats, suggesting that the induction of c-*fos* immunoreactivity in the intermediate NTS during expression of CTA is correlated with neuronal processes specific to acquisition and expression of CTA (Houpt *et al.* 1996).

7.6 Other methods

Expression of other immediate early genes besides c-*fos* has not been well documented in CTA studies. These genes include members of the *fos* (*fos*B), *jun* (c-*jun*, *jun* B, *jun* D) and *Krox* (*Krox*-20, *Krox*-24 also termed *zif*/268) families. Lanteri-Minet *et al.* (1993) observed differential temporal and spatial expression of Fos, Jun and *Krox*-24 proteins in spinal and hindbrain structures after visceral stimulation in the rat. This finding suggests that new insights into brain mechanisms of CTA learning will be obtained by using a large array of immediate early genes. Among them, *Krox*-24 should be worth trying since this gene can presumably be expressed as a manifestation of learning including long-term potentiation and synaptic plasticity (Cole *et al.* 1989).

A recent study by Lamprecht and Dudai (1995) has used the *in situ* hybridization technique to investigate the modulation of expression of the immediate early genes c-*fos*, *fos*-B, *zif*/268 and CREM in rat brain following oral administration of saccharin and i.p. injection of LiCl. c-*Fos* and *fos*-B are members of the *fos* family and exhibit a 70% homology; they form functional dimers with members of the *jun* family to produce the AP-1 transcription factors. *zif*/268 encodes a DNA-binding transcription factor with a zinc finger that binds the consensus DNA element. The cAMP responsive element modulator, CREM, is a member of the CRE-binding protein family. According to their results, i.p. injection of LiCl modulated within 40 min the expression of the mRNA of c-*fos*, *zif*/268 and CREM, but not *fos*-B, in the NTS, PBN, hypothalamic paraventricular nucleus, and central nucleus of the amygdala. The level of mRNA returned to basal level within 4 h. Saccharin CS in CTA training, however, had no detectable effect. If taste aversion learning is established on the basis of long-term potentiation in neurons of the gustatory system at some level of neuraxis, microtubule-associated protein-2 (MAP-2) may be a candidate for mediating the alterations in synaptic efficacy and dendritic remodelling that accompany learning (Johnson and Jope 1992). This structural protein is extremely susceptible to proteolysis by the calcium-activated protease, calpain (Johnson and Jope 1992). NMDA receptor activation induces dephosphorylation of MAP-2, conceivably linking this cytoskeletal protein to long-term potentiation. Woolf *et al.* (1994) have shown that Pavlovian auditory conditioning alters MAP-2 in the temporal cortex of the rat, but no such studies have been reported in the context of CTA.

One of the biochemical measures of cell activation is the use of the 2-deoxyglucose (2DG) autoradiographic technique for identification of brain regions showing enhanced glucose metabolism. Kossut and Rose (1984) examined the pattern of 2DG accumulation in the brains of chicks, and found that hyperstriatum ventrale, paleostriatum augmentatum, and lobus paraolfactorius showed significantly enhanced labelling in the birds which had been trained to avoid a bead covered with the bitter-tasting methylanthranilate in comparison with chicks which had pecked at the water-coated bead. Such metabolic mapping has not yet been used in studies of taste aversion learning in mammals.

7.7 Intracellular events

The above-mentioned approaches to detection of anatomical or functional markers of activation of neurons reflect intracellular events. However, little is known about the molecular events that realize the acquisition, consolidation, and retention of the representation of CTA in the brain areas. Recently, direct or indirect measures of some aspects of the underlying intracellular cascades have been studied during CTA learning.

The importance of protein synthesis in the insular gustatory cortex for the acquisition of CTA was suggested by Rosenblum *et al.* (1993) who showed that application of the protein synthesis inhibitor anisomycin into the rat gustatory cortex before and during training impaired acquisition of CTA to saccharin. Injection of anisomycin into the gustatory cortex immediately before the CS pre-exposure also attenuated latent inhibition.

In a subsequent paper, Rosenblum *et al.* (1995) showed that there was a rapid and marked increase in tyrosine phosphorylation of a set of proteins in the insular cortex but not in other brain areas after CTA training. A major protein so modulated was abundant in a membrane fraction and remained modulated for more than an hour after training. The novel taste alone induced only a small modulation of these proteins whereas administration of the malaise-inducing drug has no effect. It is known that tyrosine phosphorylation is a major signal transduction pathway involved in cellular metabolism, growth, and differentiation (Schlessinger and Ullrich 1992), and that tyrosine phosphorylation also plays a key function in neuronal activity and plasticity.

Increasing evidence suggests that the phosphorylation of specific proteins is involved in memory formation. Changes in the phosphorylation of a 47-kDa protein have been reported following learning and the induction of long-term potentiation. This particular protein appears to be a substrate for the enzyme PKC (Akers and Routtenberg 1985; Akers *et al.* 1986), and changes in PKC activity have been observed with different learning tasks and with the induction of long-term potentiation.

Krivanek (1993) measured the PKC activity in the cytosol and particulate fractions of the rat PBN, and showed that the translocation of the enzyme took place after i.p. administration of the CTA-inducing agents LiCl and $CuSO_4$. More recently, Krivanek (1996) found an elevation of the PKC level in the PBN after acquisition of CTA and showed that this increase started about 24 h after CS–US pairing, became maximum at 48 h, and diminished 120 h thereafter. Bielavska and Krivanek (1994) tested the contribution of PKC to the acquisition of CTA by injection of three PKC inhibitors, polymyxin B, H7, and staurosporine, into the PBN of the rat. They showed that only polymyxin B prevented CTA acquisition, and concluded that this blockade by polymyxin B was due to some non-specific mechanism not necessarily involving inhibition of PKC since two other more specific PKC inhibitors did not affect CTA learning. Taking these findings together, he suggests that PKC in the PBN may participate in consolidation of the CTA memory trace rather than in the CTA acquisition process.

On the other hand, the importance of PKC in CTA acquisition has been demonstrated for the insular gustatory cortex and amygdala in rats (Yasoshima and Yamamoto 1997). They injected the PKC inhibitors, polymyxin B or H7, into the gustatory insular cortex, thalamic taste relay, or amygdala at various time windows of the CTA paradigm. As shown in Fig. 7.7, rats injected with PKC inhibitors between the CS–US interval into the insular cortex and amygdala, but not into the thalamus, failed to acquire CTA. Injections of PKC inhibitors 4 h after the US presentation or just before the retention test elicited no disruptive effect. Their study shows that neuronal PKC activity in the cortex and amygdala has a key role in the acquisition phase of CTA, but not in the retrieval phase.

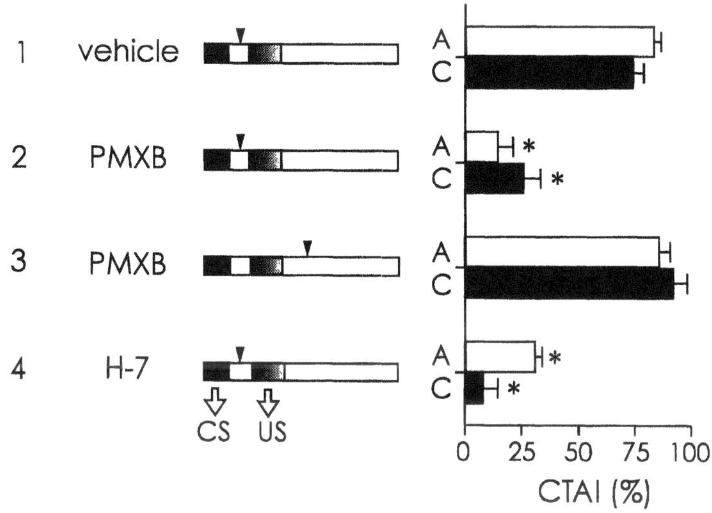

Fig. 7.7 Effects of intra-amygdalar or intracortical injections of inhibitors of protein kinase C on CTA formation. Wistar male rats received an i.p. injection of 0.15 M LiCl (20 ml/kg) as a US 60 min after free ingestion (20 min) of 0.005 M Na-saccharin (CS). Rats in groups 1, 2, 3, and 4 received 2 μl (1 μl each side) of vehicle between CS and US, 20 mM polymyxin B (PMXB) between CS and US, 20 mM PMXB after US (4 h after CS), and 20 mM H7 between CS and US, respectively, as indicated by arrow heads into either the basolateral amygdala (A) or insular cortical gustatory area (C). Both inhibitors injected into either the amygdala or gustatory cortex during the CS–US interval impaired CTA formation, but had no effect when injected after US administration. The strength of CTA formation is expressed by the CTA index (CTAI) with the formula: CTAI = 100 × (1−(sum of CS intake in 4 test sessions)/(4 × preconditioning water intake)). *Significant at $P < 0.05$ relative to vehicle controls. Modified from Yasoshima *et al.* (1997).

Since the pioneering studies by García and his colleagues (García *et al.* 1955, 1966; García and Koelling 1966) taste aversion learning has been almost exclusively the domain of psychologists and later also physiologists. In recent years, however, growing numbers of researchers with different backgrounds such as molecular biology, biochemistry, pharmacology, genetics, or immunology have been participating in CTA studies. This trend is apparently due to the unique properties of this learning which are considered attractive and useful by researchers interested in neural, cellular, or molecular mechanisms of learning and memory. It can be expected that the recent findings reviewed in this chapter will be soon followed by further intracellular, molecular, genetic, and immunohistochemical studies which may contribute decisively to the enigma of taste aversion learning.

8 Transplantation studies

8.1 Restoration of function by fetal brain grafts

The fetal brain transplant technique has been widely used to produce functional recovery from brain injury in brains of adult mammals. Thus, it has been established that the transplanted neurons can differentiate and make connections with the host brain (see Björklund and Stenevi 1985; Dunnett and Richards 1990). In several laboratories it has been demonstrated that fetal brain transplants can produce significant amelioration of cognitive deficits produced by brain lesions (for review see Dunnett and Björklund 1994). For that purpose, several learning paradigms, including simple maze learning and operant tasks, have been used to evaluate the influence of neural transplants on recovery of cognitive deficits (Dunnett and Björklund 1994). Several years ago, Bermúdez-Rattoni *et al.* (1987*b*) showed that impairment of CTA learning due to IC lesions can be restored with homotopic cortical fetal brain transplants. In brief, they made bilateral electrolytic lesions encompassing the gustatory neocortex in one group of rats, another group received bilateral lesions of the amygdala complex, and finally a third group served as unoperated control. After 15 days of post-operative recovery the animals were deprived of water for 24 h and trained to drink water daily during 5-min trials for 10 days. The consumption volume was taken every day. On the acquisition day, 0.1% saccharin was presented as a CS and followed 30 min later by intragastric LiCl (0.15 M, 190 mg/kg) as the US. Two extinction trials, during which the saccharin CS was presented, alternated with two water intake baseline measures. Volumes of liquids consumed were taken on each day.

Two days after the second extinction trial, the lesioned animals were randomly divided into four groups. Half of the cortical lesioned animals received homotopical fetal cortical grafts and half of the amygdala lesioned animals received amygdala fetal tissue. The remaining half of each lesioned group served as control to determine the effects of the lesion *per se*. The tissue grafts were prepared as follows: 17-day-old fetuses were removed from the abdominal cavity of pregnant rats under anaesthesia. The brain was extracted and the temporoparietal area (above the rhinal sulcus) was dissected under a microscope to obtain the insular cortex, while the temporal lobe was dissected to obtain the fetal amygdala. Blocks of tissue (about 3 mm^3) were then stereotaxically placed with the aid of a Hamilton microsyringe (50 μl) into the lesioned area according to the stereotaxic co-ordinates used when making the lesions (Fig. 8.1). After 8 weeks of recovery,

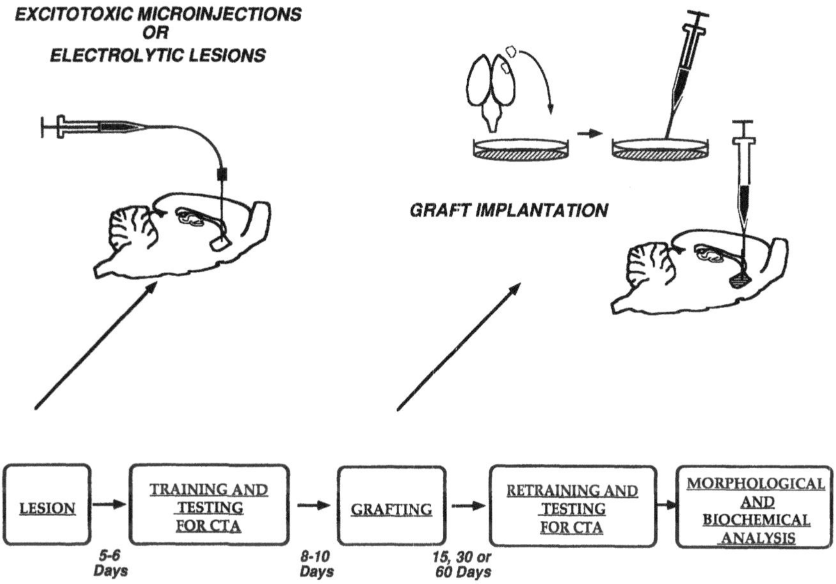

Fig. 8.1 Schematic representation of the experimental protocol. Excitotoxic or electrolytic lesions of the IC. Training and testing for CTA. Grafting procedure: the tissue from embryonic donor rats is obtained; the solid block is put into a Petri dish and aspirated by a 50 μl Hamilton syringe. CTA retraining and testing morphological and biochemical analysis of brain tissue.

the five groups (the controls, the cortical and amygdala grafted groups, and the cortical and amygdala lesioned animals) were tested with the behavioural procedure described above. However, one additional acquisition trial was given to all groups.

Results indicated that lesioned animals tested before transplantation showed the expected disrupted taste aversion when compared with the unoperated controls (Fig. 8.2A). After transplantation animals with cortical grafts recovered the ability to learn the taste–illness association and were not significantly different from the unoperated controls (Fig. 8.2B). The possibility of spontaneous recovery was excluded, because the animals with IC lesions that did not receive transplants were unable to acquire CTA 8 weeks after the transplantation procedure even after two acquisition trials (Bermúdez-Rattoni *et al.* 1987*b*). Similar results were reported by Yirmiya *et al.* (1988) who used saccharin as a conditioned stimulus for training before transplantation and saline as the CS after brain grafting. The results were similar, that is rats with cortical implants, but not lesioned controls, recovered the ability to learn the CTA tasks several weeks after transplantation but the recovery induced by grafts was not complete.

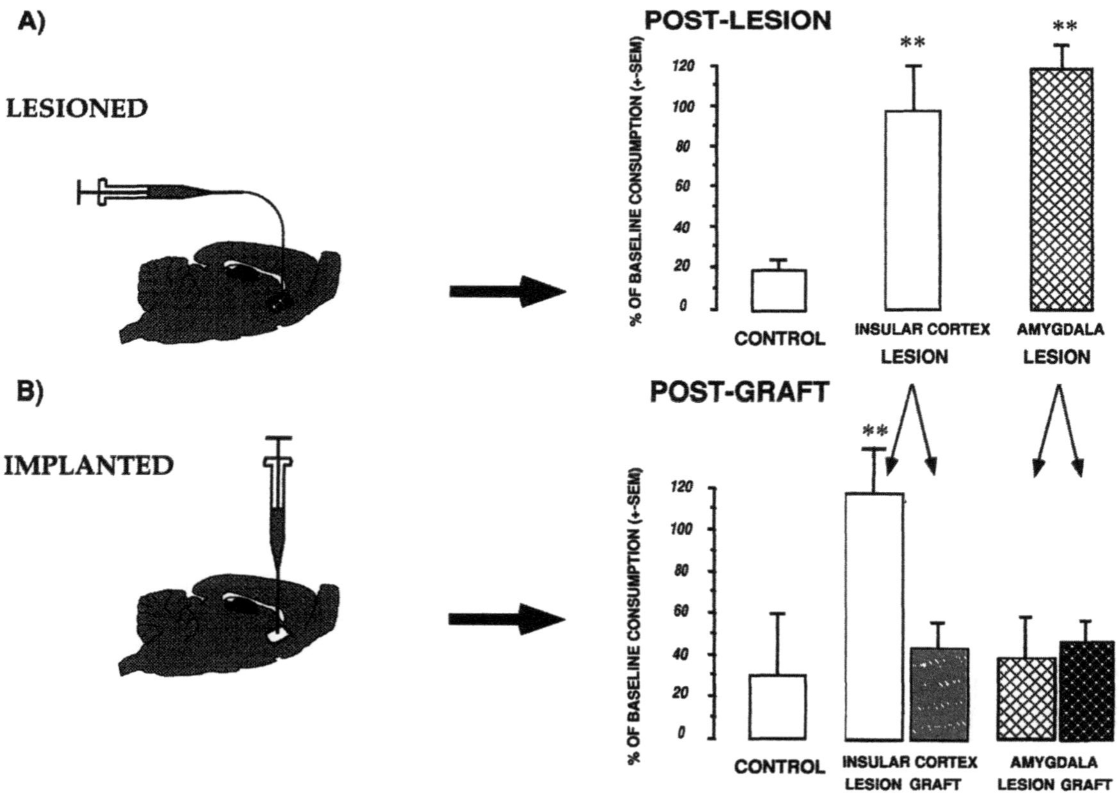

Fig. 8.2 The effect of cortical and amygdala lesions and subsequent fetal brain grafts on CTA acquisition. The bars illustrate the amount of the CS fluid consumed by controls, cortical or amygdala lesioned animals (A) and by controls, grafted or lesioned animals 8 weeks after transplantation (B). Consumption is expressed as the percentage of each group's water baseline on the previous day. $^{**}P < 0.01$ (Dunnett test).

Surprisingly, Bermúdez-Rattoni *et al.* (1987*b*) found that non-grafted animals with damaged amygdala showed spontaneous recovery 8 weeks after the lesion (Fig. 8.2B). Spontaneous recovery of performance in different behavioural tasks has been reported in animals tested 6 weeks after large cortical ablations (Dunnett and Björklund 1987). The functional differences between the amygdala and the IC have been discussed in detail elsewhere (Bermúdez-Rattoni *et al.* 1987*a*, 1989). One possible explanation is that, for taste aversion learning, IC may be a permanent memory storage site, whereas amygdala may only facilitate an initial stage of CTA storage. This interpretation is supported by the fact that NMDA lesions of amygdala sparing the 'fibres of passage' do not affect the acquisition of CTA, whereas similar lesions of the IC have strong disruptive effects on CTA (Aggleton *et al.* 1981; Bermúdez-Rattoni and McGaugh 1991; see Chapter 3 for review).

The first conclusion that could be drawn from these findings was that the recovery of function seen in animals with cortical grafts might be due to the formation of connections between the graft tissue and the host brain. Several experiments have shown the existence of reciprocal connections between the thalamus and cortical grafts (Sharp and Gonzalez 1986). Therefore, it was decided to test the putative connections between the IC grafts and the host tissue. An earlier series of experiments carried out in our laboratory showed that the degree of functional recovery induced by fetal brain tissue depends on the place from which the graft was taken. That is, the animals that received homotopic insular cortical tissue recovered the CTA; the animals that received either occipital or tectal heterotopic tissue or no transplant showed no behavioural recovery.

Therefore, in another series of experiments animals that received either heterotopical or homotopical grafts after behavioural testing were studied with HRP histochemistry, to see if the transplants were able to re-establish connectivity with areas that have connections with the IC in the intact brain (Escobar *et al.* 1989). The authors followed the horseradish peroxidase protocol according to Mesulam (1982) and counterstained the sections with thionine. The slices were examined under bright and dark field microscopy for the presence and location of retrogradely labelled neurons. In addition, some control animals, IC, and tectal grafted animals received a unilateral injection (0.5 μl) of HRP into the right amygdala (Escobar *et al.* 1989).

Sections from control animals which received HRP into the amygdala showed reaction product boundaries extending 1 mm in diameter around the area of the tracer application. HRP labelled cells were always found in the ipsilateral gustatory neocortex (Krieg's area 13 and 14), and in the ventromedial nucleus of the thalamus (see Chapter 3). In general, the insular brain transplants appeared to be healthy and localized in the appropriate target area of the host brain. In the insular cortical brain transplants we found only few HRP labelled cells, although a large amount of them appeared in the VPM and VM thalamic nuclei in the same animals. These results indicate that the brain grafts could establish connections with the host brain. The low density found in the grafts has been previously described in studies that used HRP for tracing the host–graft projections. Thus, few labelled cells have been found in fetal brain grafts in the hippocampus and in the occipital and somatosensory cortex (Björklund and Stenevi 1979; Kromer *et al.* 1980; Jaeger and Lund 1981; Ross and Ebner 1985). The reasons for the low density of labelled cells were discussed elsewhere (Jones 1975; Escobar *et al.* 1989). It is necessary to stress that the fetal brain transplant is still developing and its neurons are starting to make connections with the host tissue. It is possible, therefore, that the transplant–host connections have not yet reached a complete mature form. On the other hand, heterotopic tectal transplants did not integrate well with the host. There was a heavy glial invasion, necrosis, abundant vacuoles, and scant vascularization. No HRP labelled cells were found in the grafted tissue. Nevertheless, abundant HRP labelled cells in the VM and VPM nuclei of the thalamus of the same rats demonstrated the effectiveness of the HRP technique used (Escobar *et al.* 1989).

8.2 Cholinergic mechanisms in functional recovery

All the experiments discussed so far were made with long (60-day) waiting times between grafting and post-graft retraining. Fernández-Ruíz *et al.* (1991) attempted to establish the best post-graft time for observation of functional recovery. For that purpose, they used rats with IC lesions disrupting taste aversion. The animals received cortical transplants and were retrained at 15, 30, 45, and 60 days after transplantation. The behavioural results showed almost complete functional recovery at 60 days, slight recovery at 45 and 30 days, and poor recovery at 15 days post-graft. In addition, HRP injections into the amygdala or VPM thalamus revealed that, at 15 days, there were no HRP labelled cells in the grafted tissue. At 30, 45, and 60 days post-graft, increasing numbers of connections with the thalamus and with the amygdala were found in the graft, though not as many as seen in the controls. The behavioural recovery was correlated with increased histochemically detected acetylcholinesterase activity of the graft and with morphological maturation, revealed by Golgi staining. These results suggest that some morphological recovery was necessary for the acquisition of taste aversion.

Analysis of the functional activity of brain grafts should identify some neurochemical elements that could be responsible for functional recovery. Since the cholinergic neurotransmitter system is believed to play an important role in learning and memory processes (Decker and McGaugh

1991), López-García *et al.* (1990) made biochemical assays of cholinergic activity present in brain grafts and compared it with control levels in the same regions of intact brain. The first study demonstrated that slices taken from the IC are able to release labelled GABA, ACh, and glutamate, but not dopamine. Additionally, significant amounts of glutamic acid decarboxylase, choline acetyltransferase, and acetylcholinesterase activity were found in intact IC homogenates (López-García *et al.* 1990). Biochemical analyses showed that IC fetal grafts released GABA, ACh, and glutamate in response to potassium depolarization, whereas occipital grafts released labelled GABA and glutamate, but not ACh. These results suggest that cholinergic transmission is important for CTA learning and that ACh may play a role in graft mediated behavioural recovery.

The implication of this series of grafting experiments was that functional recovery is related to the morphological maturation and/or cholinergic activity of the graft. It has been suggested by other investigators, however, that the structural and morphological integrity of fetal brain grafts might not be essential for behavioural recovery after cortical brain injury. These investigators have speculated that both brain injury and brain grafts induce a release of neurotrophic factors that can reactivate neural function and/or prevent injury-induced degeneration in the damaged host brain. Moreover, Dunnett and Björklund (1987) have reported that neocortical grafts produce short-term improvement in the T-maze alternation performance, which can be attributed to diffuse influences of the embryonic tissue on the lesioned host brain, rather than to reconnection of the damaged circuits. In addition, Labbe *et al.* (1983) showed that cortical grafts could restore alternation learning deficits in rats with large aspiration lesions of the prefrontal cortex. The rats with cortical grafts were able to learn a T-maze alternation task more rapidly than rats with lesions alone or with heterotopic grafts from embryonic cerebellar tissue. These authors used the HRP technique to demonstrate that functional grafts established connections with the host brain. However, behavioural recovery was seen 7 days after the transplantation whereas the anatomical connectivity was analysed several months later. Therefore, the functional recovery induced by the graft could be attributed to a trophic factor rather than to restored anatomical connectivity (Dunnett and Björklund 1994).

In the light of the above findings, research into restoration of CTA learning after lesions of insular cortex should examine the alternative hypothesis by testing the effects of pharmacological agents facilitating behavioural recovery after brain injury, i.e. of neurotrophic factors, gangliosides, etc. Several experiments reported that nerve growth factor (NGF) produces significant regeneration, regrowth, and penetration of cholinergic axons into the hippocampal formation (Gage 1990; Hagg *et al.* 1990). It is also known that lesions disrupting cholinergic innervation of the fimbria fornix produce severe impairment of different learning tasks and that constant infusion of NGF into the lesioned brain has produced functional and anatomical recovery (Varon *et al.* 1989). Moreover, aged rats with functional impairments that received constant intracerebral infusions of NGF improved enormously after several weeks of treatment (Gage and Björklund 1986).

8.3 The role of nerve growth factor in recovery

As mentioned above, exogenous application of NGF can enhance and accelerate the regrowth and penetration of cholinergic axons into previously lesioned fimbria fornix (Hagg *et al.* 1990; Vahlsing *et al.* 1991). This treatment restores rhythmic theta activity in the denervated hippocampus (Buzsaki *et al.* 1987), and stimulates the recovery of cholinergic functions (Lapchak *et al.* 1991). The finding that IC, but not occipital fetal grafts, releases ACh in response to potassium depolarization (López-García *et al.* 1990) suggests that cholinergic transmission may play a role in graft-mediated behavioural recovery.

Recently, Bermúdez-Rattoni *et al.* (1992) evaluated the role of NGF in the graft induced recovery of the ability of IC lesioned rats to acquire CTA 15, 30, and 60 days after transplantation. Several groups of rats in which CTA learning was disrupted by electrolytic lesions of IC were grafted as follows: one group received gelfoam plus NGF, another received fetal IC graft plus NGF, the third group received gelfoam plus vehicle and the fourth group received only fetal IC graft. An additional group of intact animals was used as an unoperated control. All groups were subdivided into three subgroups that were retrained for CTA at 15, 30, and 60 days post-graft, respectively. The control groups had strong taste aversions at all three intervals. The IC graft plus NGF group showed a significantly recovered ability to acquire the taste aversions at the three post-graft times when compared with the control group. The groups receiving the IC graft without NGF showed a disrupted taste aversion 15 days after transplantation but recovered the ability to acquire CTA when tested 30 and 60 days after grafting. The groups that received gelfoam plus NGF or gelfoam plus vehicle remained significantly impaired at all three post-graft times tested, when compared with the control group. The results indicate that the administration of NGF to animals with cortical grafts accelerates their ability to acquire the CTA at 15 days post-graft (Bermúdez-Rattoni *et al.* 1992).

Since previous studies showed that in the absence of NGF, IC grafts only start to induce functional recovery 30 days after implantation, NGF appears to accelerate recovery. In other studies, Escobar *et al.* (1993, 1994) tested the possibility that the combination of NGF with other fetal brain tissue could produce similar functional recovery. Groups of IC-lesioned animals showing

CTA disruption received insular or occipital cortical grafts plus NGF, IC grafts plus vehicle, or remained lesioned as a control group. The results showed that the combination of NGF with insular but not with occipital cortical grafts restored the ability to acquire CTA at 15 days post-graft (Escobar *et al.* 1994). In an earlier study, Escobar *et al.* (1993) demonstrated that choline acetyltransferase (ChAT) activity in the IC grafts with but not without NGF was similar to the normal IC activity of unoperated controls at 15 days post-graft. In addition, glutamic acid decarboxylase activity in the same groups of animals changed similarly as the ChAT activity. It was concluded that the NGF added to the IC grafts induced recovery of learning abilities correlating with the recovery of ChAT activity (Escobar *et al.* 1993). Recent experiments were designed to investigate the involvement of the cholinergic neurotransmitter system in behavioural recovery induced by the fetal brain grafts in combination with NGF. For that purpose, Russell *et al.* (1994) used *in vivo* assays of ACh turnover in the IC of rats subjected to lesion and implant treatments similar to those already described: IC graft + NGF, IC graft + vehicle, occipital graft + NGF, NGF alone, and lesioned animals. The biochemical results showed that IC implants combined with NGF were at non-lesioned control levels of ACh. In contrast, animals that received heterotopic occipital grafts with NGF or received NGF alone showed significantly higher levels of ACh and deficits in learning and memory when compared with controls. The results suggested that there was some up-regulation of the cholinergic function that accompanies IC lesions, and that this can be reversed, like the behavioural deficits, by IC implants supplemented with NGF (Russell *et al.* 1994).

In order to further assess the role of cholinergic activity of the IC grafts in behavioural recovery, rats showing disrupted taste aversions due to IC lesions were grafted with homotopical IC grafts supplemented with NGF. Eight days later all animals were implanted with a cannula above each insular cortex. After 15 days of recovery the animals were subdivided into two groups. One received microinjections of scopolamine (a muscarinic cholinergic antagonist) 15 min before acquisition of CTA and the other group received microinjections of saline as control. One week later the groups were reversed and received either scopolamine or saline before the acquisition of the Morris water-maze task. The results showed that the groups receiving scopolamine, but not those receiving saline, showed significant disruption of both CTA and water-maze learning. These results support the assumption that functional cholinergic blockade of insular grafts impairs acquisition of two different learning tasks (Bermúdez-Rattoni *et al.* 1995).

8.4 Recovery of learning capabilities

The experiments described above showed that it is possible to restore acquisition of CTA by cortical fetal brain grafts. Heterotopic fetal grafts did not produce the recovery of function seen with the use of homotopic grafts. In other models of cortical lesions, it has been shown that learning deficits could be improved by heterotopic brain grafts or even by implantation of glial elements (Nieto-Sampedro and Cotman 1986). In the IC grafting model, the best recovery is seen 30 to 40 days after the transplantation, i.e. at a time when the maturity of the grafts is fully expressed in morphological and biochemical assays (Escobar *et al.* 1989; Fernández-Ruíz *et al.* 1991). Other studies showing that IC grafts, but not occipital fetal grafts, release Ach in response to K^+ depolarization suggested that cholinergic transmission may play a role in the graft-mediated behavioural recovery (López-García *et al.* 1990). It is noteworthy that functional recovery in IC lesioned animals can be achieved by homotopic grafting, whereas heterotopic grafts (occipital cortex) do not produce functional recuperation (Bermúdez-Rattoni *et al.* 1989; Escobar *et al.* 1994). Finally, it was shown that tissue samples of the fetal brain transplants without NGF addition taken at various intervals after grafting are at different stages of maturation. That is, at 15 days there was little development of neurons and blood vessels in the implanted tissue. By 60 days, greater neuronal density developed: abundant vascularization and proliferation of glial cells was apparent, and increasing numbers of fibres crossed the border between the graft and the host tissue (Fernández-Ruíz *et al.* 1991). Together, these results suggested that the behavioural effects are related to integration and maturation of the implanted tissue.

These studies clearly support the view that there are functional differences between homotopical vs. heterotopical grafts. Despite the fact that functional recovery can be accelerated by the supplementation of the homotopic graft with NGF, the results are essentially the same. That is, the accelerated recovery induced by supplemented NGF is only seen with the homotopically but not with the heterotopically grafted embryonic cortices. In this regard, Castro *et al.* (1991) showed that homotopic but not heterotopic grafted somatosensory cortices produced full recovery of electrical responses in the host cortex. Moreover, these authors showed that neocortical tissue grafted into the cortex of newborn rats can develop interconnections with the host central nervous system including transplant efferents to the host cortex, thalamus, striatum, and spinal cord (Floeter and Jones 1984, 1985; Castro *et al.* 1985, 1988, 1989). In addition, the graft received afferents from several areas of the host brain including the ipsi- and contralateral cortices, thalamus, and hypothalamus (Chang *et al.* 1984, 1986; Castro *et al.* 1988). Cortical grafts were also found to ameliorate thalamic atrophy induced by cortical lesion (Sharp and González 1986). The neocortical grafts placed into the somatosensory cortex cavities of adult rats were responsive to stimulation of vibrissae and to tactile stimuli (Bragin *et al.* 1988). Castro *et al.* (1991) grafted either homotopic (parietal) or heterotopic (occipital) cortical tissue in aspiration lesions of the somatosensory cortex of newborn rats. After 2.5 to 5 months, they started electrophysiological experiments examining whether thalamic afferents to the grafts are functionally integrated with the host's somatosensory pathway, and relay peripheral sensory information to the transplant. The results showed that both homotopical and heterotopical grafts responded to thalamic stimulation. Surprisingly, stimulation on the contralateral forepaw produced cortical responses in the homotopically but not in the heterotopically grafted tissue (Fig. 8.3). Moreover, the heterotopic transplants did not ameliorate thalamic atrophy induced by cortical lesions and assessed by volumetric measurements (Castro *et al.* 1991). Despite the fact that both grafts sent projections to the host spinomedullary junction, they differed by their electrically evoked responses to the same stimuli. It was concluded that heterotopic grafts produce apparent anatomical recovery that is, however, not accompanied by complete functional recovery (Neafsey *et al.* 1988).

Finally, the restorative effects of brain grafts supplemented with NGF can be blocked by local application of scopolamine in two different learning tasks, CTA and the Morris water maze. These results suggest that restoration is implemented by cholinergic activity of the fetal brain grafts. A possible dual role for cholinergic activity in the morphological recovery and in the behavioural recovery has yet to be determined. In this regard, transient expression of acetylcholinesterase activity during early development has been shown in some thalamocortical projection systems (Robertson *et al.* 1988, 1991). These results suggest that the cholinergic system may contribute not only to normal neural development and to the differentiation of the thalamocortical projections, but also to the development of connectivity of cortical grafts.

8.5 Fetal brain grafts and locus of CTA engram

The above results show that IC is involved in the acquisition and retention of different learning tasks. Nevertheless, our experiments challenge the hypothesis that the IC is a permanent memory store for aversively motivated learning. Therefore, we used the fetal brain grafts as a 'reversible lesion' technique, to further assess the role of the IC in CTA storage.

A question important in this context is whether the IC is involved only in the CTA and inhibitory avoidance learning (acquisition and consolidation), or whether it also plays a role in long-term storage of these memories (retention and retrieval). To test the long-term memory hypothesis Ormsby

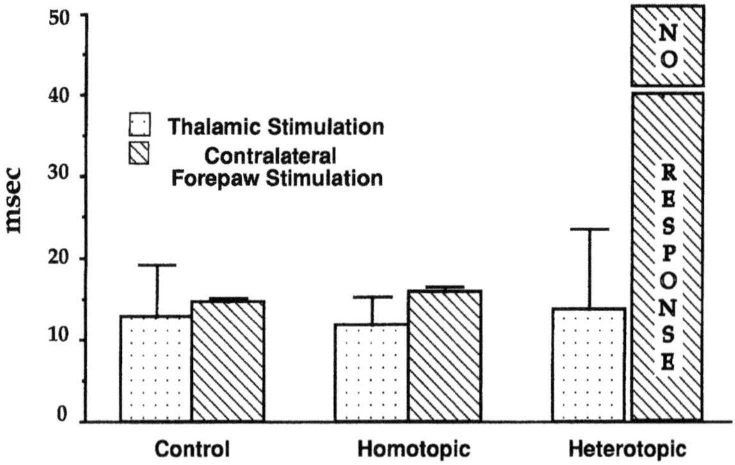

Fig. 8.3 Distribution of orthodromic latencies of single units recorded in control, sensorimotor (homotopic) and occipital (heterotopic) cortex grafts to electrical stimulation of the ipsilateral ventral thalamus and contralateral forepaw. Modified from Neafsey *et al.* (1989) and Castro *et al.* (1991).

et al. (1991) used three groups of intact rats that were trained in CTA and subsequently in inhibitory avoidance. One group received bilateral NMDA-induced lesions of the IC 4 days after CTA training to allow for consolidation. The other two groups used as controls received sham operations or remained unoperated. After a week of recovery the animals were given the test trials for taste aversion. The results showed that the IC lesions induced a significant impairment of CTA retrieval indicating that the IC is involved in long-term memory, as well as in the learning processes.

To further specify the role that the IC might play in either retention and retrieval or in both phases of CTA memory, Ormsby *et al.* (1994) performed another series of studies. Taking into account the findings of Bermúdez-Rattoni *et al.* (1987*b*, 1989), Escobar *et al.* (1989, 1994), and Fernández-Ruíz *et al.* (1991), in which homotopic fetal IC grafts induced recovery of the ability to learn CTA 45 days after transplantation, the authors examined the possibility that the grafts could also produce recovery of the memory processes. To achieve this, they elicited CTA in four groups of intact rats; 4 days later three groups received lesions induced by bilateral microinjections of NMDA. This procedure impairs acquisition of CTA and inhibitory avoidance (Bermúdez-Rattoni and McGaugh 1991). The fourth group served as a non-lesioned control. Eight days after surgery, one group received homotopical grafts, one group received heterotopical occipital grafts and the last group was used as lesioned control. As expected, the non-lesioned controls had a good recall of the CTA but, surprisingly, both grafted groups (those with homotopical or heterotopical grafts) also recovered the ability to recall (Fig. 8.4a). None of the lesioned controls showed recall of the task. When the animals were retrained for a second CTA and in an inhibitory avoidance task, only the homotopically grafted group was able to learn both tasks (Fig. 8.4b). It is noteworthy that the rats with the occipital heterotopic grafts did recall, but did not recover the ability to acquire a new CTA using a different taste stimulus or the inhibitory avoidance. The group grafted with homotopic IC grafts recovered the ability to acquire this last task, but the lesioned group did not. These results corroborate the notion that the IC is involved in the learning and retrieval of CTA, but probably is not the site of long term storage.

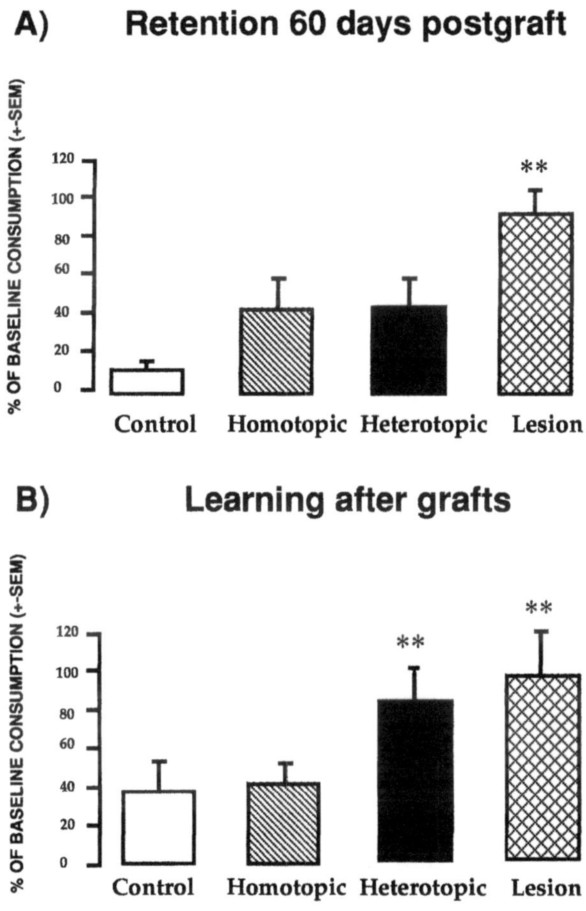

Fig. 8.4 Effects of post-training IC grafts on CTA. (A) Effects on retrieval of CTA acquired before the pre-lesion. Saccharin consumption is expressed as a percentage of each group's previous day water baseline. (B) Effects on acquisition of a different CTA (with saline solution as CS).

When investigating the role of the IC in recall, we have used the fetal brain grafts to change a permanent lesion into a 'reversible lesion' thus making the IC again available during the recall processes. It appears that under normal circumstances the IC is important for the acquisition and consolidation of aversively motivated learning tasks, but that it is less important for the long-term recall of taste aversion. When fetal brain grafts provide some 'reversibility' to surgical lesions, it is possible to decide whether the studied structure is a real memory storage site or only an exit place for information stored elsewhere in the brain. Research reviewed in Chapter 4 demonstrated that interaction between one hemicortex and the ipsilateral subcortical parabrachial nucleus is necessary for the formation and retrieval of CTA (Gallo and Bures 1991). Our results together with those reviewed in Chapter 4 suggest that IC is an important structure for the acquisition and consolidation of CTA, but that other cortical and subcortical structures are involved in CTA retrieval. In other words, it is conceivable that a functional loop established between the IC and the parabrachial nucleus is responsible for CTA retrieval.

An alternative hypothesis currently evaluated in our laboratory suggests that cholinergic activity is necessary for acquisition of new but not for retrieval of old taste information already stored somewhere in the brain. In this regard, Toumane and Durkin (1993) showed that cortical cholinergic activity is particularly important in a specific time window of the post-learning period. Cholinergic markers measured in the cortex showed a decreasing gradient of activity after the animals had learned a concurrent spatial discrimination task and were tested 30 s to 15 min later (Durkin and Toumane 1992). It is conceivable that the insular grafts but not the occipital grafts have enough cholinergic activity to support functional recovery in the IC lesioned animals (Escobar *et al.* 1994; Russell *et al.* 1994; Bermúdez-Rattoni *et al.* 1995; Ormsby *et al.* 1994). More research will be necessary to clarify the role of the IC as a permanent storage place for other aversively motivated learning tasks (e.g. for place navigation and inhibitory avoidance learning).

9 Paradoxes, projections, and perspectives of CTA research

Since its discovery (García *et al.* 1955) CTA was accepted with some distrust because it did not fit the traditional models of animal learning. CTA violated the putative equivalence of stimuli, i.e. the assumption that any CS can be associated with any US, by clearly showing that the taste of food is a more salient CS than its colour, texture, or shape in the food aversion or food preference context. Still more suspicious was the hour-long delay between CS and US which contrasted with the few seconds typical for other types of conditioning. Attempts to explain this delay by CS–US contiguity mediated by lingering tastes or by regurgitation of ingesta (Bitterman 1976) elicited heated discussion (Revusky 1977) but were disproved by experiments demonstrating CTA learning even in animals drinking masking solutions in the CS–US interval (Kalat and Rozin 1971) or consuming a rapidly degrading substance (e.g. sucrose, Revusky 1968) or fluids of high or low temperature (Nachman 1970) as the CS.

Almost fifty years of research have convincingly demonstrated that CTA is a highly specialized, dedicated form of learning associating signals generated during consumption of food with the delayed visceral consequences of its ingestion. CTA research played an important role in the renaissance of learning theory which accepted learning as an adaptive process and abandoned rigid definitions based on generalizations of laboratory abstractions. The updated learning theory became more tolerant to phenomena irreconcilable with earlier views. Rather than stressing CTA peculiarities the present investigators attempt to detect the presence of some features of CTA learning, albeit on a different time scale, in the more conventional forms of learning. CTA remains, nevertheless, a challenging phenomenon, relevant to a number of hot spots of contemporary neurobiology and neuropsychology.

9.1 CTA and consciousness

CTA can be considered an extension of the unconditioned reflexes, protecting the homeostasis of internal environment against the effect of ingested toxins. Whereas the innate reactions reject bitter, acid, or irritating diets, CTA makes it possible to avoid any flavoured diet which was followed within a definite time window by intoxication. The underlying memory is often acquired by implicit learning. According to Bernstein (1985) the cause of most CTAs in humans cannot be traced back to the acquisition day. Absence of episodic memories recording the events leading to CTA formation is consonant with the well-documented observation that strong CTA is formed even when poisoning occurred under deep anaesthesia or other comatose states (see Chapter 1). This aspect was stressed by García (1990) who considers CTA a unique example of effective integration of conscious and unconscious streams of information. When ionizing radiation is used as the US, the rats can recognize its presence as an innocuous odour to which they rapidly habituate. After the cumulative effects of radiation exposure become noxious, the animal does not attempt to avoid radiation but rather to modify its feeding preferences. García (1990) postulates the existence of a toxiphobic mind system which selectively attaches the nauseous visceral state to any preceding salient taste stimulus even when the rat is not aware of the nausea and does not react to it by any unconditioned or conditioned responses.

In some respects, CTA resembles the phenomenon of blindsight (Weiskrantz 1986), the capability of patients with post-geniculate lesions of the optic radiation to emit reflexive responses and even voluntarily initiated responses to visual stimuli although they claim to see nothing. The effective processing of visual information contrasts with the absent conscious perception of visual phenomena. It is more difficult to determine whether animals are aware of events eliciting overt responses (Weiskrantz 1995), although several methods have been suggested (Dickinson 1988; Stoerig and Cowey 1995) to reveal conscious avoidance of the presented stimuli.

Although CTA is not the only example of learning without awareness (Bures *et al.* 1988), it is the only instance of such learning which has adaptive significance and gives the animal a clear advantage in the struggle for survival. This is the reason why CTA is particularly well suited for research into the mechanism of learning without awareness, which not only has important theoretical implications but may perhaps open new approaches to treatment of drug abuse and control of pathological emotional states. It is conceivable that similar disruption-resistant stages of memory formation also contribute to other forms of memory but that their short duration prevented their identification with the available functional ablation procedures.

9.2 Limited role of proteosynthesis

Many studies have shown that protein synthesis inhibitors (PSIs) produce impairment in the formation of long-term memory (Davis and Squire 1984). A paradoxical aspect of taste aversion learning is that PSIs such as anisomycin and cycloheximide can act as US to induce a CTA when administered intraperitoneally or subcutaneously (Booth and Simson 1973; Nakajima 1974; Squire *et al.* 1975). According to Squire *et al.* (1975), anisomycin or cycloheximide produced similar CTA as LiCl at concentrations disrupting retention of a shock avoidance learning. Thus, the widely accepted assumption that all forms of learning share the same cellular and molecular mechanisms of plasticity contrasts with the possibility to elicit CTA by doses of PSIs which disrupt most other forms of learning. PSIs, or antibiotics, are included in the list of drugs that elicit CTAs (Gamzu *et al.* 1985). Since peripheral injection of PSIs often produces observable sickness in animals (Nakajima 1974), there is no doubt that PSIs themselves exert US effects in the CTA paradigm. The problem is how effectively and in what time course peripherally injected PSIs inhibit protein synthesis in the brain sites responsible for CTA formation because it is well established that direct infusions of PSIs into the brain impair CTAs to the CS paired with the US such as LiCl (Tucker and Gibbs 1976, 1979; Tucker and Oei 1982; Rosenblum *et al.* 1993). Tucker and Gibbs (1976, 1979) and Tucker and Oei (1982) investigated the relationship between amnesia for saccharin aversion and inhibition of protein synthesis elicited by cycloheximide in rats, and suggested a time window in which memory can be disrupted by this PSI. Rosenblum *et al.* (1993) showed that application of anisomycin into the rat gustatory cortex impaired CTA acquisition (1) when it elicited in the target region more than 90% inhibition of protein synthesis lasting for more than 90 min, (2) when the injection was made a few minutes before saccharin ingestion, and (3) when the drug was infused exactly into the gustatory cortex, i.e. delivery 2 mm above the insular cortex, did not affect CTA. These findings suggest that hour-long nearly complete inhibition of protein synthesis in the relevant site within a limited time window is necessary for PSIs to exert their effects. It is doubtful that such strict criteria can be attained with peripheral injections of PSIs (Patterson *et al.* 1989). Even if systemically administered PSIs really inhibit protein synthesis in the brain, it is possible that CTA formation requiring protein synthesis for long-term gustatory memory can outlast the pharmacological action of PSIs in the case that their aversive US effects last longer than their inhibitory effects. An obvious experiment yet to be done is to monitor c-*fos* expression in the brain after peripheral administration of PSIs. c-*Fos* expression in the lateral PBN after peripheral application of PSIs would confirm that PSIs have exerted only US effects. A real paradox would be complete absence of c-*fos* expression in the brain including lateral PBN after peripheral application of PSIs dosages eliciting CTA.

9.3 CTA produced by rewarding drugs

CTA can be induced by a wide spectrum of various agents. Emetic mechanisms appear to be involved in the formation of CTAs produced by many of them, but there is contrary evidence for other agents, suggesting that some CTAs are mediated by emetic mechanisms and others are not (Grant 1987). LiCl belongs to the former, while psychoactive (rewarding) drugs which are known to act as positive reinforcers such as morphine, amphetamine, and cocaine may belong to the latter. These drugs are known to induce CTAs within a dose range similar to that self-administered by animals. It has also been known that many rewarding drugs that produce CTA also produce a conditioned place preference at equivalent dosages. This finding is often referred to as a 'paradox'.

CTAs induced by rewarding drugs such as morphine appear to be qualitatively distinct from CTAs induced by emetic agents (Hunt and Amit 1987; Parker 1995). In contrast to emetic CTAs, CTAs induced by rewarding drugs are generally weaker, show a greater between-subject variability of the taste aversion response, and are more susceptible to disruption by drug pre-exposure. CTAs induced by rewarding drugs appear to exhibit a curvilinear dose–response curve, in contrast to the linear dose–response function seen in emetic CTAs. Intraoral infusion of sucrose solution previously paired with LiCl elicits aversive responses such as gaping, chin rubbing, and paw treading, which are not elicited by sucrose paired with rewarding drugs. Lesions of the area postrema attenuate LiCl-induced CTAs, but not amphetamine-induced CTAs (Chapter 3), suggesting that LiCl-paired sucrose elicits conditioned sickness. Rats bury the spout of a bottle containing LiCl-paired or unconditionally aversive (e.g. quinine) tastes, but not that containing amphetamine-paired milk, although they avoid drinking it.

It is apparent from the above and other evidence that CTAs produced by rewarding and emetic drugs are different. Rats do not avoid sucrose paired with rewarding drugs because it tastes bad, the neural mechanism of such flavour avoidance is still unclear. It has been suggested that the avoidance may be motivated by fear (Pelchat *et al.* 1983) or by taste shyness (Hunt and Amit 1987) or that each drug produces its own specific US properties for association (Pelchat *et al.* 1983). Alternatively, if a single common neural substrate is responsible for the formation of taste aversion/avoidance learning, a wide variety of USs including rewarding drugs could activate common brain sites. This assumption can be tested by the c-*fos* immunohistochemical technique. According to Swank *et al.* (1995) (Fig. 7.2), both LiCl and amphetamine injections induce elevations in c-*fos*-like immunoreactivity (c-FLI) relative to saline-injected controls. The most dramatic elevations were seen in intermediate NTS where the c-FLI response to LiCl was more than threefold greater than that elicited by amphetamine. In the lateral PBN amphetamine and LiCl induced comparable increases in c-FLI positive cells relative to saline controls. Increases in c-FLI after amphetamine and LiCl were also seen in area postrema, but the intensity of the reaction was not remarkable. Amphetamine induced dense concentrations of c-FLI in striatum while no c-FLI was visible here after LiCl. These results indicate that the lateral PBN is a site of common and comparable activation by these two very different drugs. This observation is consistent with accumulating evidence for a key role of the PBN in acquisition of CTAs (see Chapters 3, 4, and 7) induced by a variety of pharmacological agents including LiCl and amphetamine (Sakai and Yamamoto 1997). The finding that both rewarding and emetic agents activate the lateral part of PBN is particularly important in the light of the assumption that this region is a crucial site for the CS–US association as well as a projection site for negative hedonics and quality of US. According to Stolerman and Kumar (1972), experienced opiate users use the magnitude of the initial nausea occurring following drug intake as a positively correlated cue for predicting the intensity of a subsequent drug 'high'. This report infers that rewarding agents have initial dysphoric (aversive) and later euphoric effects.

Anatomical dissociation of rewarding and aversive effects of morphine has been studied by microinjections of this drug into confined brain sites. It was concluded that the medial forebrain bundle pathway mediates the rewarding effects while the PBN and the nucleus accumbens mediate the aversive effects (Jaeger and van der Kooy 1993). Bechara *et al.* (1993) showed that bilateral ibotenic acid lesions of the lateral PBN blocked CTA induced by morphine but not conditioned place preference induced by this drug. This suggests that the lateral PBN is a critical link in the neural pathway carrying the aversive motivational effects of opiates from the gut into the central nervous system, independent of the neural pathway carrying the rewarding motivational effects of morphine.

Taken together, both emetic and rewarding drugs may elicit two types of reactions: emetic drugs activate the emetic centre and the hedonically aversive site located in the lateral PBN, while rewarding drugs activate the same hedonically aversive site and also the euphoric reward system. In contrast to emetic drugs, rewarding drugs induce both positive and aversive feelings, thus leading to taste avoidance (fear and/or taste shyness) reflecting such conflicting feelings. Since the aversive action of rewarding drugs is often not clearly expressed, combined application of emetic agents with rewarding drugs might increase the aversive effect. It is conceivable that this approach may be exploited in drug abuse treatment.

It must be pointed out, however, that the above explanation of CTA induced by drugs of abuse is not generally accepted. According to a recent alternative explanation proposed by Grigson (1997), psychoactive drugs decrease CS consumption not by their aversive properties but only by their rewarding effects. Intake of the gustatory CS is reduced not only when it precedes an aversive visceral state but also when it predicts access to a highly rewarding US. This so-called anticipatory contrast occurs when access to moderately palatable saccharin solution is followed by availability of more preferred sucrose (Flaherty *et al.* 1994). According to the reward comparison hypothesis CS intake is reduced because the hedonic value of the CS is outweighed by the expected availability of the highly reinforcing US, represented by another more palatable taste or by the drug-induced pleasure. While anticipatory contrast may certainly produce taste avoidance closely similar to that induced by cocaine or morphine, the hypothesis is not easily applicable to situations when the CS serves also as the US (e.g. in the case of alcohol) and when its consumption provides access to subsequent pleasure.

9.4 Molecular and cellular nature of CTA plasticity

Learning and memory, which are the highest forms of adaptation, are reflected by plastic changes in brain function and behaviour. Efficient learning, robust memory, and delayed associability of the CS and US make CTA a useful system for analysing molecular, cellular, and neuroanatomical mechanisms of acquisition, association, consolidation, retention, and retrieval of memories. Pharmacological studies of the molecular and cellular mechanisms implementing taste aversion learning should employ agonists (or activators) and antagonists (or inhibitors) with well-known pharmacological effects. Affecting various stages of CTA formation by microinfusions of such drugs into critical brain sites elaborates the results obtained with non-specific functional ablations using TTX or local anaesthetics as well as with the classical electrolytic or excitotoxic brain lesions. That is, the reversible effects of substances with well-known mechanisms of action make it possible to elucidate not only the neural circuits involved but also the membrane receptors and intracellular messengers relevant for CTA formation.

The underlying process may have a pharmacology identical to that of hippocampal long-term potentiation (Izquierdo 1994). Pharmacological analysis of the inhibitory avoidance task (Izquierdo 1993; Wolfman *et al.* 1994; Bernabeu *et al.* 1995; Izquierdo *et al.* 1995) indicates an early involvement of glutamatergic NMDA, AMPA, and metabotropic receptors regulated by CO, NO, and platelet-activating factor and by GABAergic and cholinergic transmission. This is followed in the next 1–2 h by a phase sensitive to Ca^{2+}/calmodulin-dependent protein kinase II and PKC inhibitor and still later by a phase sensitive to protein synthesis inhibitors. The expression of memory is blocked by AMPA (but not NMDA) receptor antagonists applied into the hippocampus, amygdala, or entorhinal cortex up to 20–26 days after training. It is not yet known, however, how similar or different is the cascade of molecular and cellular processes involved in CTAs. The first result pointing in this direction is the blockade of CTA acquisition by microinjection of muscarinic antagonists into the insular cortex shortly before saccharin drinking but not in the CS–US interval (Naor and Dudai 1996) or of PKC inhibitors into the PBN shortly after the CS–US association (Bielavska and Krivanek 1994).

Besides the pharmacological studies, cellular/molecular observations following CTAs can be performed by immunohistochemical techniques. For example, the muscarinic–cholinergic system and the closely associated PKC might be visualized with the monoclonal antibodies raised against purified muscarinic acetylcholine receptors and PKC g and ab isomer proteins. Considering the fact that CTA is robust and maintained for many hours or days, it is also important to know the molecular processes required for the morphological changes in neuronal connectivity which presumably encode memory. Since GAP-43 protein has been shown to be essential for axonal growth (Colley and Routtenberg 1993), it may play a central role in CTA formation by propelling the growth of new axons and terminals or the modification of existing presynaptic terminals. Alterations in GAP-43 protein can be examined immunohistochemically, i.e. increased or decreased staining may be observed in brain regions possibly participating in CTA formation.

The experiments reviewed in Chapter 7, showing that the gustatory CS and malaise-inducing US elicit modulation of the expression of immediate-early genes including c-*fos* are in agreement with the molecular hypothesis asserting that the encoding of long-term memory traces involves modulation of neuronal gene expression (Davis and Squire 1984; Dudai 1989, 1996). According to this hypothesis, the stimuli that trigger the encoding of long-term memory initiate molecular cascades that modulate immediate-early genes, which in turn modulate the expression of late-response genes, leading to long-lasting alterations of the neuronal circuits that encode the relevant stimuli.

The above hypothesis should be verified by deciding whether the expression of immediate-early genes, especially c-*fos*, in brain regions participating in CTA formation is a prerequisite of the encoding of CTA memory or is merely an epiphenomenon correlated with CTA training. First relevant results were obtained by Lamprecht and Dudai (1996) who found that phosphorothioate-modified oligodeoxynucleotides antisense to c-*fos* microinjected into rat amygdala several hours before CTA training impaired taste aversion memory tested 3–5 days after conditioning. Swank *et al.* (1996) also reported that c-*fos* antisense, infused into the fourth ventricle of mice, reduced LiCl-induced c-*fos* translation in NTS and PBN neurons and blocked both acquisition and extinction of CTA without interfering with the sensory processing of either CS or US. The authors suggested that c-*fos* antisense may block association of CS and US encoded by inducible bZIP proteins. It can be expected that the role of other immediate-early genes and late-response genes in encoding CTA will soon be elucidated by local application of specific antisense oligodeoxynucleotides into the brain regions showing increased expression of these genes during various phases of CTA acquisition and retrieval.

Finally, the importance of genetic studies will increase in the future. The separation of CTA-prone and CTA-resistant rats (Elkins *et al.* 1992) may reveal the existence of specific genes linked to the formation of CTAs. The specificity of such deficit can be analysed by producing the gene-deficient animals by the gene-targeting method and by comparing the efficiency of such animals in various learning tasks.

9.5 Electrical substitutes for natural CS and US

One of the characteristic features of CTA is highly efficient learning: strong CTA can be induced by a single pairing of CS and US. Electrophysiological results (Matsuo and Yamamoto 1990; Matsuo *et al.* 1994; Niijima and Yamamoto 1994) show that impulse discharges of the afferent nerves in response to gustatory and gastrointestinal stimuli continue for more than several scores of seconds and several scores of minutes, respectively. This long-lasting excitation of the CS and US pathways represents a major difference between CTA and other types of conditioning supported by stimuli such as electric shock, sound, and light which last for only a brief time (usually less than 1 s).

One of the disadvantages of the CTA paradigm is that the onset and offset of the chemical stimuli used as the CS and US cannot be determined as exactly as in the case of the above-mentioned physical stimuli. It is conceivable that with similarly short duration of the gustatory CS and illness-inducing US, CTA would be acquired only after repeated CS–US pairing like classical conditioning. Since it is difficult to control the duration of afferent discharges elicited by the natural CS and US stimuli, substitution by electrical stimulation would be the best way to test this possibility. Gallo *et al.* (1988) investigated the feasibility of substituting electrical intracerebral stimulation of the area postrema for the aversive US in a CTA paradigm, and found that single pairing of strawberry or coconut extracts with 0.1 ms rectangular pulses of 50 Hz, delivered for 4 h after a 15-min delay following ingestion of the gustatory CS, produced reliable CTA learning. Anodal DC current applied to the tongue surface elicits a unique taste called electric taste in humans and also elicits good responses in the taste nerve of the rat only during the stimulation period (Bujas *et al.* 1979; Yamamoto *et al.* 1980*b*). This method could be used as a substitute for the conventional gustatory CS. In fact, Weijnen (1977) paired a low anodal current applied to the tongue with LiCl poisoning and obtained CTA to this electrical CS.

A similar approach was used by Phillips and Le Piane (1978): when rats receiving electrical stimulation of the basolateral nucleus of the amygdala while drinking plain tap water were injected with LiCl, they greatly reduced their water intake in a retention test. This result suggests that brain stimulation can serve as a CS in a CTA paradigm.

The idea of replacing the gustatory CS by electrical current was further elaborated by Greenshaw *et al.* (1985) who used lick-triggered rewarding stimulation of the medial forebrain bundle as the CS followed by LiCl poisoning. The assumption that rewarding brain stimulation contingent upon licking can be interpreted as a kind of pleasant taste was supported by the finding that two self-stimulation–LiCl pairings reduced the lick rate to 40% of the preconditioning level. The fact that the same stimulus triggered by nose poking could not be associated with poisoning indicates that the CS is not the electrical current alone but only when applied in the context of consummatory behaviour.

The possibility to induce CTA by pairing (or pairings) of two electrical stimulations, e.g. anodal pulses applied to the tongue and intracranial stimulation of the area postrema, would prove that they can substitute the natural CS and US. Pairing electrical stimuli of different duration at different CS–US intervals will make it possible to find conditions optimizing the use of these stimuli for establishing CTA. Electrical CS and US could also be useful in electrophysiological studies of the neural substrate of CTA attempting to identify the association site of the CS and US, the storage site of the gustatory short-term memory of the CS and the nature of the synaptic modifications implementing the underlying plastic change.

9.6 Conditioned immunity

For decades, it was believed that the immune system is independent and autonomous. However, accumulating evidence indicated the existence of a relationship between the central nervous system and the immune system. It has been demonstrated that stress and/or depression impair immune responses to diverse immunological challenges. In this regard, several reports in the literature have related personality factors, stress situations, and experimental induction of sickness in man, with immunological suppression (Maier *et al.* 1994; Merril and Jonakait 1995). Further, neural systems are important modulators of immune responses; these include the hypothalamic–pituitary–adrenal axis (Blalock 1994), the autonomic nervous system (Felten *et al.* 1987), and the limbic system (Ader *et al.* 1995; Pan and Long 1993). There are reports showing the limbic structures and the autonomic nervous system receive immune inputs (Abreu *et al.* 1994; Yu and Shinnick-Gallagher 1994). It has been demonstrated that hypothalamic lesions or stimulation of the hypothalamus induce changes in both cellular and humoral immunological responses. Further, immunological responses induced significant increase of electrical activity in the ventromedial nucleus of the hypothalamus that matches with increased immunological activity after antigenic challenge (Cross *et al.* 1980; Roszman *et al.* 1982).

Therefore, it is not surprising that immune responses can be modulated by behavioural conditioning procedures (Ader and Cohen 1981; Alvarez-Borda *et al.* 1995). In the early 1970s Ader and Cohen (1975) accidentally found that CTA can be induced when using cyclophosphamide as the unconditioned stimulus. After several exposures to the aversively conditioned taste stimulus many experimental animals died. This did not happen when a different unconditioned stimulus was used in similar experiments. Besides producing CTA due to gastrointestinal distress, cyclophosphamide also has well-known immunosuppressive effects (Ader and Cohen 1982). By using immunological techniques, Ader *et al.* (1995) discovered that presentation of the taste stimulus to the conditioned animals elicits not only strong CTA but also strong immunosuppression which may account for increased mortality.

Research reviewed in Chapter 3 showed the important role of the insular cortex in the neural mediation of CTA. Taking advantage of the CTA model of conditioned immunosuppression, Ramírez-Amaya *et al.* (1996) demonstrated that NMDA-induced lesions of the insular cortex disrupt the acquisition of conditioned immune responses and CTA. The authors used three groups of rats. Baseline antibody titres were obtained from all animals by enzyme-linked immunosorbent assay (ELISA) to estimate the anti-ovalbumin (OVA) IgG production. Three days later, two different groups of animals received NMDA lesions of either insular or parietal cortices, and a third group received sham insular cortex lesion as control. Two additional groups were non-lesioned controls: one was a conditioned control and the other was a control group in which the CS was not presented during the test (see Fig. 9.1).

After recovery from surgery all groups received one single pairing of saccharin with an i.p. injection of cyclophosphamide. Twenty-five days later all rats were immunized with OVA (500 μg, in a volume of 300 μl) and 30 min later were re-exposed to the taste of saccharin. Blood samples were taken 4, 8, and 12 days after immunization. It is noteworthy that at the time of immunization the immunosuppressive effects of cyclophosphamide have fully disappeared. The results showed that lesions of the insular cortex but not of the parietal cortex disrupted the acquisition of both conditioned immunosuppression and CTA. Rats with insular cortex lesions showed similar antibody production as the group to which conditioned stimulus was not presented (see Fig. 9.1). Furthermore, the authors were able to demonstrate in other experiments that the insular cortex lesions interfered neither with the normal humoral immune response nor with the immunosuppressive effects of cyclophosphamide. It was concluded that the insular cortex could play an important role in the associative mechanisms of immunosuppressive conditioning (Ramírez-Amaya *et al.* 1996).

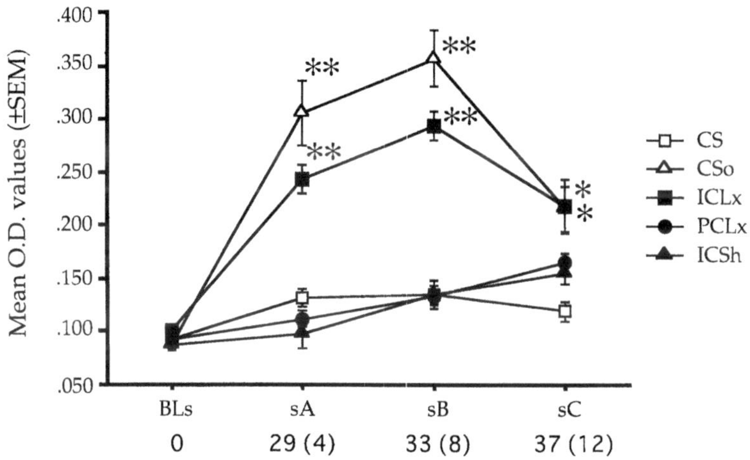

Fig. 9.1 Mean optic density values (± standard error of the mean) obtained from ELISA represent IgM production in conditioned animals with or without lesions. BLs is the baseline sample taken at the beginning of the experiment. Samples sA, sB, and sC were taken 29, 33, and 37 days after pairing saccharin with an i.p. injection of cyclophosphamide. Numbers in parentheses indicate days after immunization with ovalbumin and re-exposure to saccharin. Groups: CS, conditioned; CSo, conditioned but not re-exposed to saccharin; ICLx, insular cortex lesion; PCLx, parietal cortex lesion; ICSh, insular cortex sham. *$P < 0.05$; **$P < 0.01$. From Ramírez-Amaya *et al.* (1996).

Whereas several groups studied conditioned suppression of immune responses, conditioned enhancement of antibody production has been investigated for the first time by Alvarez-Borda *et al.* (1995). These authors paired the novel taste of saccharin with injection of a protein antigen (egg-white lysozyme) and found after re-exposure to the conditioned taste a reliable increase of antibody production. The conditioned production of antibodies of IgG and IgM classes proceeded in a pattern resembling in both magnitude and temporal course a normal secondary response after re-injection of the same dose of the antigen (Alvarez-Borda *et al.* 1995). The above experiments illustrate the impact of the CTA paradigm upon a new and fascinating area of neuroimmunology.

References

Abrams, T. W. and Kandel, E. R. (1988). Is contiguity detection in classical conditioning a system or a cellular property? Learning in Aplysia suggests a possible molecular site. *Trends in Neuroscience, 11,* 128–135.

Abreu, P., Llorene, E., Hernández, M. M., and González, M. C. (1994). Interleukin-1b stimulates tyrosine hydroxylase activity in the median eminence. *NeuroReport, 5,* 1356–1358.

Adachi, A. (1984). Projection of the hepatic vagal nerve in the medulla oblongata. *Journal of Autonomic Nervous System, 10,* 287–293.

Adachi, A. (1994). Mechanisms of fluid homeostasis mediated by the brainstem. In I. R. A. Barraco (Ed.), *Nucleus of the Solitary Tract* (pp. 399–405). Boca Raton: CRC Press.

Adachi, A. and Kobashi, M. (1988). Electrophysiological analysis of chemosensitive neurons within the area postrema of the rat. *Progress in Brain Research, 74,* 77–83.

Adachi, A., Kobashi, M., Miyoshi, N., and Tsukamoto, G. (1991). Chemosensitive neurons in the area postrema of the rat and their possible functions. *Brain Research Bulletin, 26,* 137–140.

Ader, R. (1976). Conditioned adrenocortical steroid elevations in the rat. *Journal of Comparative and Physiological Psychology, 90,* 1156–1163.

Ader, R. (1977). A note on the role of olfaction in taste aversion learning. *Bulletin of the Psychonomic Society, 10,* 402–404.

Ader, R. and Cohen, N. (1975). Behavioral conditioned immunosuppression. *Psychosomatic Medicine, 37,* 333–340.

Ader, R. and Cohen, N. (1981). Conditioned immunopharmacological responses. In R. Ader (Ed.), *Psychoneuroimmunology* (pp. 185–228). New York: Academic Press.

Ader, R. and Cohen, N. (1982). Behavioral conditioned immunosuppression and murine systemic lupus erythematosus. *Science, 215,* 1534–1536.

Ader, R., Grota, L. J., and Buckland, R. (1978). Effects of adrenalectomy on taste aversion learning. *Physiological Psychology, 6,* 359–361.

Ader, R., Cohen, N., and Felten, D. (1995) Psychoneuroimmunology: interactions between the nervous system and the immune system, *Lancet, 345,* 99–103.

Aggleton, J. P., Petrides, M., and Iversen, S. D. (1981). Differential effects of amygdaloid lesions on conditioned taste aversion learning by rats. *Physiology and Behavior, 27,* 397–400.

Aguado, L., San Antonio, A., Perez, L., Del Valle, R., and Gomez, J. (1994). Effects of the NMDA receptor antagonist ketamine on flavor memory: conditioned aversion, latent inhibition, and habituation of neophobia. *Behavioral and Neural Biology, 61,* 271–281.

Aguero, A., Arnedo, M., Gallo, M., and Puerto, A. (1993*a*). The functional relevance of the lateral parabrachial nucleus in lithium chloride-induced aversion learning. *Pharmacology, Biochemistry and Behavior, 45,* 973–978.
Aguero, A., Arnedo, M., Gallo, M., and Puerto, A. (1993*b*). Lesions of the lateral parabrachial nuclei disrupt aversion learning induced by electrical stimulation of the area postrema. *Brain Research Bulletin, 30,* 585–592.
Aguero A., Gallo, M., Arnedo, M., Molina, F, and Puerto, A. (1996). Effects of lesions of the medial parabrachial nucleus (PBNm): taste discrimination and lithium chloride-induced aversion learning after delayed and contiguous interstimulus intervals. *Psychobiology, 24,* 265–280.
Aguero A., Gallo, M., Arnedo, M., Molina, F, and Puerto, A. (1997). The functional relevance of medial parabrachial nucleus in intragastric sodium chloride-induced short-term (concurrent) aversion learning. *Neurobiology of Learning and Memory, 67,* 161–166.
Akers, R. F. and Routtenberg, A. (1985). Protein kinase C phosphorylates a 47 Mr protein (F1) directly related to synaptic plasticity. *Brain Research, 334,* 147–151.
Akers, R. F., Lovinger, D. M., Colley, P. A., Linden, D. J., and Routtenberg, A. (1986). Translocation of protein kinase C activity may mediate hippocampal long-term potentiation. *Science, 231,* 587–589.
Albert, M. and Ayres, J. J. B. (1989). With number of preexposures constant, latent inhibition increases with prexposure CS duration or total CS exposure. *Learning and Motivation, 20,* 278–294.
Aleksanyan, Z. A., Buresova, O., and Bures, J. (1976). Modification of unit responses to gustatory stimuli by conditioned taste aversion in rats. *Physiology and Behavior, 17,* 173–179.
Allen, W. F. (1941). Effects of ablating the piriform amygdaloid areas and hippocampus on positive and negative olfactory conditioned reflexes and on conditioned olfactory differentiation. *American Journal of Physiology, 132,* 81–92.
Alvarez-Borda, B., Ramírez-Amaya, Pérez-Montfort, R., and Bermúdez-Rattoni, F. (1995). Enhancement of antibody production by a learning paradigm. *Neurobiology of Learning and Memory, 64,* 103–105.
Amit, Z., Levitan, D. E., Brown, Z. W., and Rogan, F. (1977). Possible involvement of central factors in the mediation of conditioned taste aversion. *Neuropharmacology, 16,* 121–124.
Anderson, C. E., Tilson, H. A., and Mitchell, C. L. (1982). Conditioned taste aversion following acutely administered acrylamide. *Neurobehavioral Toxicology and Teratology, 4,* 497–499.
Archer, T. and Sjoden, P.-O. (1979). Positive correlation between pre- and postconditioning saccharin intake in taste-aversion learning. *Animal Learning and Behavior, 7,* 144–148.
Arnedo, M., Gallo, M., Aguero, A., and Puerto, A. (1990). Effects of medullary afferent vagal axotomy and area postrema lesions on short-term and long-term NaCl-induced taste aversion learning. *Physiology and Behavior, 47,* 1067–1074.

Arnedo, M., Gallo, M., Aguero, A., and Puerto, A. (1991). Differential effects of subdiaphragmatic vagotomy on NaCl-induced aversion learning. *Behavioral and Neural Biology, 55,* 141–153.

Arnedo, M., Gallo, M., Aguero, A., and Puerto, A. (1993). Medullary afferentvagal axotomy disrupts NaCl-induced short-term taste aversion learning. *Behavioral and Neural Biology, 59,* 69–75.

Ashe, J. H. and Nachman, M. (1980). Neural mechanisms in taste aversion learning. *Progress in Psychobiology and Physiological Psychology, 9,* 233–262.

Baker, T. B. and Cannon, D.S. (1979). Alcohol and taste-mediated learning. *Addictive Behaviors, 7,* 221–230.

Bakner, L., Strohen, K., Nordeen, M., and Riccio, D. C. (1991). Postconditioning recovery from the latent inhibition effect in conditioned taste aversion. *Physiology and Behavior, 50,* 1269–1272.

Barker, L. M. and Smith, J. C. (1974). A comparison of taste aversions induced by radiation and lithium chloride in CS–US and US–CS paradigms. *Journal of Comparative and Physiological Psychology, 87,* 644–654.

Bartoshuk, L. M. (1979). Bitter taste of saccharin related to the genetic ability to taste the bitter substance 6-*N*-propylthiouracil. *Science, 205,* 934–935.

Batsell, W. R., Jr., and Best, M. R. (1992). Variations in the retention of taste aversions: evidence for retrieval competition. *Animal Learning and Behavior, 20,* 146–159.

Batsell, W. R., Jr., and Best, M. R. (1993). One bottle too many? Method of testing determines the detection of overshadowing and retention of taste aversions. *Animal Learning and Behavior, 21,* 154–158.

Battista, A. F. (1967). Effect of cold on cortical potentials in the cat. *Experimental Neurology, 19,* 140–155.

Baum, M., Foidart, D. S., and Lapointe, A. (1974). Rapid extinction of a conditioned taste aversion following unreinforced intraperitoneal injection of the fluid CS. *Physiology and Behavior, 12,* 871–873.

Baylis, L. L. and Rolls, E. T. (1991). Responses of neurons in the primate taste cortex to glutamate. *Physiology and Behavior, 49,* 973–979.

Bechara, A., Martin, G. M., Pridgar, A., and van der Kooy, D. (1993). The parabrachial nucleus: a brain substrate critical for mediating the aversive motivational effects of morphine. *Behavioral Neuroscience, 107,* 147–160.

Beckstead, R. M. and Norgren, R. (1979). An autoradiographic examination of the central distribution of the trigeminal, facial, glossopharyngeal, and vagal nerves in the monkey. *Journal of Comparative Neurology, 184,* 455–472.

Beckstead, R. M., Morse, J. R., and Norgren, R. (1980). The nucleus of the solitary tract in the monkey: projections to the thalamus and brain stem nuclei. *Journal of Comparative Neurology, 190,* 259–282.

Bellingham, W. P. and Lloyd, D. (1987). Injected flavor as a CS in the conditioned aversion preparation. *Animal Learning and Behavior, 15,* 62–68.

Benjamin, R. M. and Akert, K. (1959). Cortical and thalamic areas involved in taste discrimination in the albino rat. *Journal of Comparative Neurology, 111,* 231–259.

Berenbaum, M. R. and Miliczy, E. (1983). Mantids and milkweed bugs: efficacy of

aposematic coloration against invertebrate predators. *American Midland Naturalist, 111,* 64–68.
Berger, B. D. (1970). Learning in the anesthetized rat. *Federation Proceedings, 29,* 749.
Berger, B. D. (1972). Conditioning of food aversions by injections of psychoactive drugs. *Journal of Comparative and Physiological Psychology, 81,* 21–26.
Berger, B. D., Wise, C. D., and Stein, L. (1973). Area postrema damage and bait shyness. *Journal of Comparative and Physiological Psychology, 82,* 475–479.
Bermúdez-Rattoni, F. and McGaugh, J. L. (1991). Insular cortex and amygdala lesions differentially affect acquisition of inhibitory avoidance and conditioned taste aversions. *Brain Research, 549,* 165–170.
Bermúdez-Rattoni, F., Rusiniak, K. W., and García, J. (1983). Flavor–illness aversions: potentiation of odor by taste is disrupted by application of novocaine into amygdala. *Behavioral and Neural Biology, 37,* 61–75.
Bermúdez-Rattoni, F., Grijalva, C. V., Kiefer, S. W., and García, J. (1986). Flavor–illness aversions: the role of the amygdala in the acquisition of taste-potentiated odor averions. *Physiology and Behavior, 38,* 503–508.
Bermúdez-Rattoni, F., Fernández, R. J., Chávez-Almanza, F., Coburn, K. I., and García, J. (1987*a*). Flavor–illness aversions. Potentiation of odor by taste is regulated by the cholinergic activity of amygdala and dorsal hippocampus. *Pharmacology, Biochemistry and Behavior, 26,* 553–559.
Bermúdez-Rattoni, F., Fernández, J., Sánchez, M. A., Aguilar Roblero, R., and Drucker-Colín, R. (1987*b*). Fetal brain transplants induce recuperation of taste aversion learning. *Brain Research, 416,* 147–152.
Bermúdez-Rattoni, F., Forthman, D. L., Sanchez, M. A., Perez, J. L., and García, J. (1988). Odor and taste aversions conditioned in anethetized rats. *Behavioral Neuroscience, 102,* 726–732.
Bermúdez-Rattoni, F., Fernández, J., and Escobar, M. L. (1989). Fetal brain transplants induce recovery of morphological and learning deficits of cortical lesioned rats. In L. E. Cañedo, L. E. Todd, L. Packer, and J. Jaz (Eds.), *Cell function and disease* (pp. 261–273). New York: Plenum Press.
Bermúdez-Rattoni, F., Escobar, M. L., Piña, A. L., Tapia, R., López-García, J. C., and Hiriart, M. (1992). Effects of NGF on the recovery of conditioned taste aversion in insular cortex lesioned rats. In R. L. Doty (Ed.), *Chemical signals in vertebrates* (pp. 297–303). New York: Plenum Press.
Bermúdez-Rattoni, F., Ormsby, Ch., Escobar, M. L., and Hernández-Echeagaray E. (1995). The role of the insular cortex in the acquisition and long lasting memory for aversively motivated behavior. In J. L. McGaugh, F. Bermúdez-Rattoni, and R. A. Prado-Alcalá (Eds.), *Plasticity in the central nervous system: learning and memory* (pp. 67–82). New Jersey: Erlbaum.
Bernabeu, R., Princ, F., Levi de Stein, M., Fin, C., Juknat, A. A., Batile, A., Izquierdo, I., and Medina, J. H. (1995). Evidence for the involvement of hippocampal CO production in the acquisition and consolidation of inhibitory avoidance learning. *NeuroReport, 6,* 516–518.
Bernstein, I. L. (1978). Learned taste aversions in children receiving chemotherapy. *Science, 200,* 1302–1303.

Bernstein, I. L. (1985). Learned food aversions in the progression of cancer and its treatment. *Annals of the New York Academy of Sciences, 443,* 365–380.

Bernstein, I. L. and Fenner, D. P. (1983). Learned food aversions: heterogeneity of animal models of tumor-induced anorexia. *Appetite, Journal for Intake Research, 4,* 79–86.

Bernstein, I. L. and Webster, M. M. (1980). Learned taste aversions in humans. *Physiology and Behavior, 25,* 363–366.

Bernstein, I. L., Vitiello, M. V., and Sigmundi, R. A. (1980). Effects of tumor growth on taste-aversion learning produced by antitumor drugs in the rat. *Physiological Psychology, 8,* 51–55.

Berridge, K., Grill, H. J., and Norgren, R. (1981). Relation of consummatory responses and preabsorptive insulin release to palatability and learned taste aversions. *Journal of Comparative of Physiological Psychology, 95,* 363–382.

Best, M. R. (1975). Conditioned and latent inhibition in taste-aversion learning: clarifying the role of learned safety. *Journal of Experimental Psychology: Animal Behavior Processes, 1,* 97–113.

Best, M. R. and Meachum, C. L. (1986). The effects of stimulus preexposure on taste mediated environmental conditioning: potentiation and overshadowing. *Animal Learning and Behavior, 14,* 1–5.

Best, P. and Orr, J. (1973). Effects of hippocampal lesions on passive avoidance and taste aversion conditioning. *Physiology and Behavior, 10,* 193–196.

Bielavska, E. and Krivanek, J. (1994). Intracerebral injection of polymyxin B blocks the acquisition of conditioned taste aversion in rats. *Neuroscience Letters, 182,* 239–242.

Bielavska, E. and Roldan, G. (1996). Ipsilateral connections between the gustatory cortex, amygdala and parabrachial nucleus are necessary for acquisition and retrieval of conditioned taste aversion in rats. *Behavioural Brain Research, 81,* 25–31.

Bitterman, M. E. (1976). Flavor aversion studies. *Science, 192,* 266–267.

Björklund, A. and Stenevi, C. (1979). Reconstruction of the nigrostriatal dopamine pathway by intracerebral nigral transplants, *Brain Research, 177,* 555–560.

Björklund, A. and Stenevi, C. (1985). *Neural grafting in the mammalian CNS.* Amsterdam: Elsevier.

Blake, W. D. and Lin, K. K. (1978). Hepatic portal vein infusion of glucose and sodium solution and the control of drinking in the rat. *Journal of Physiology, 274,* 129–139.

Blalock, J. E. (1994). The syntax of immune-neuroendocrine communication. *Immunology Today, 15,* 504–511.

Bond, N. W. and DiGiusto, E. L. (1976) One trial higher-order conditioning of a taste aversion. *Australian Journal of Psychology, 28,* 53–55.

Boegman, R. J., Cockhill, J., Jhamandas, K., and Beninger, R. J. (1992). Excitotoxic lesions of the rat basal forebrain: differential effects on choline acetyltransferase in the cortex and amygdala. *Neuroscience, 51,* 129–135.

Bolas, K. C., Bellingham, W. P., and Martin, G. M. (1979). Aversive properties of cycloheximide versus memory inhibition in chickens' formation of visually cued food aversions. *Pharmacology, Biochemistry and Behavior, 10,* 251–254.

Booth, D. A. and Simson, P. C. (1973). Aversion to a cue acquired by its association with effects of an antibiotic in the rat. *Journal of Comparative and Physiological Psychology, 84,* 319–323.

Booth, D. A. and Simson, P. C. (1974). Taste aversion induced by a histidine-free amino acid load. *Physiological Psychology, 2,* 349–351.

Booth, D. A., Pilcher, C. W. T., D'Mello, G. D., and Stolerman, I. P. (1977). Comparative potencies of amphetamine, fenfluramine and related compounds in taste aversion experiments in rats. *British Journal of Pharmacology, 61,* 669–677.

Borison, H. L., and Wang, S. C. (1953). Physiology and pharmacology of vomiting. *Pharmacological Review, 5,* 193–200.

Bradley, R. M. and Mistretta, M. (1971). Intravascular taste in rats as demonstrated by conditioned aversion to sodium saccharin. *Journal of Comparative and Physiological Psychology, 75,* 186–189.

Bragin, A. G., Bohne, A., and Vinogradova, O. S. (1988). Transplants of the embryonal rat somatosensory neocortex in the barrel field of the adult rat: responses of the grafted neurons to sensory stimulation. *Neuroscience, 25,* 751–758.

Braun, J. and McIntosh, H., Jr. (1973). Learned taste aversions induced by rotational stimulation. *Physiological Psychology, 1,* 301–304.

Braun, J. and Snyder, D. R. (1973). Taste aversions and acute methyl mercury poisoning in rats. *Bulletin of the Psychonomic Society, 1,* 419–420.

Braun, J. J., Slick, T. B., and Lorden, J. F. (1972). Involvement of gustatory neocortex in the learning of taste aversions. *Physiology and Behavior, 9,* 637–641.

Braun, J. J., Kiefer, S. W., and Ouellet, J. V. (1981). Psychic ageusia in rats lacking gustatory neocortex. *Experimental Neurology, 95,* 363–382.

Braun, J. J., Lasiter, P. S., and Kiefer, S. W. (1982). The gustatory neocortex of the rat. *Physiological Psychology, 10,* 13–45.

Braveman, N. S. (1975). Relative salience of gustatory and visual cues in the formation of poison based food aversions by guinea pigs (*Cavia porcellus*). *Behavioral Biology, 14,* 189–199.

Brookshire, K. H., Stewart, C. N., and Bhagavan, H. N. (1972). Saccharin aversion in alloxan-diabetic rats. *Journal of Comparative and Physiological Psychology, 79,* 385–393.

Brower, L. P. and Fink, L. S. (1985). A natural toxic defense system: cardenolides in butterflies vs. birds. *Annals of the New York Academy of Sciences, 443,* 171–188.

Brown, Z. W., Amit, Z., Smith, B., and Rockman, G. E. (1978). Differential effects on conditioned taste aversion learning with peripherally and centrally administered acetaldehyde. *Neuropharmacology, 17,* 931–935.

Brozek, G., Buresova, O., and Bures, J. (1974). Effect of bilateral cortical spreading depression on the hippocampal theta activity induced by oral infusion of the aversive gustatory stimulus. *Experimental Neurology, 42,* 661–668.

Brozek, G., Buresova, O., and Bures, J. (1979*a*). Electrophysiological analysis of retrieval of conditioned taste aversion: physiological techniques and data processing. *Physiologia Bohemoslovaca, 28,* 537–544.

Brozek, G., Siegfried, B., Klimenko, V. M., and Bures, J. (1979*b*). Lick-triggered intracranial stimulation interferes with retrieval of conditioned taste aversion. *Physiology and Behavior, 23,* 625–631.

Bujas, Z., Frank, M., and Pfaffmann, C. (1979). Neural effects of electrical taste stimuli. *Sensory Processes, 3,* 353–365.

Bures, J. and Buresova, O. (1981). Elementary learning phenomena in food selection. In G. Adam, I. Meszaros, and E. I. Banyai (Eds.), *Brain and behaviour. Advances in physiological sciences, Vol. 17* (pp. 81–94). Budapest: Akademiai Kiado.

Bures, J. and Buresova, O. (1989). Conditioned taste aversion to injected flavor: differential effect of anesthesia on the formation of the gustatory trace and on its association with poisoning in rats. *Neuroscience Letters, 98,* 305–309.

Bures, J. and Buresova, O. (1990). Reversible lesions allow reinterpretation of system level studies of brain mechanisms of behavior. *Concepts in Neuroscience, 1,* 69–89.

Bures, J., Buresova, O., and Krivanek, J. (1974). *The mechanism and applications of Leao's spreading depression of electroencephalographic activity.* New York: Academic Press.

Bures, J., Buresova, O., and Krivanek, J. (1988). *Brain and behavior. Paradigms for research in neural mechanisms.* Chichester: Wiley.

Buresova, O. (1978). Neocortico-amygdalar interaction in the conditioned taste aversion in rats. *Activitas Nervosa Superior, 20,* 224–230.

Buresova, O. and Bures, J. (1973). Cortical and subcortical components of the conditioned saccharin aversion. *Physiology and Behavior, 11,* 435–439.

Buresova, O. and Bures, J. (1974). Functional decortication in the CS–US interval decreases efficiency of taste aversion learning. *Behavioral Biology, 12,* 357–364.

Buresova, O. and Bures, J. (1975). Functional decortication by cortical spreading depression does not prevent forced extinction of conditioned saccharin aversion in rats. *Journal of Comparative and Physiological Psychology, 88,* 47–52.

Buresova, O. and Bures, J. (1977). The effect of anesthesia on acquisition and extinction of conditioned taste aversion. *Behavioral Biology, 20,* 41–50.

Buresova, O. and Bures, J. (1979). Extinction of a newly acquired conditioned taste aversion: effect of gustatory CS administered under anesthesia. *Behavioural Processes, 4,* 329–339.

Buresova, O. and Bures, J. (1980). Post-ingestion interference with brain function prevents attenuation of neophobia in rats. *Behavioural Brain Research, 1,* 299–312.

Buresova, O. and Bures, J. (1981). Threshold hypothermia disrupting acquisition of conditioned taste aversion and attenuation of neophobia in rats. *Behavioral and Neural Biology, 31,* 274–282.

Buresova, O., Bures, J., Hassmanova, J., and Fifkova, E. (1964). The effect of decreased body temperature on conditined reflex activity in the rat. *Physiologia Bohemoslovaca, 13,* 220–226.

Buresova, O., Aleksanyan, Z.A., and Bures, J. (1979). Electrophysiological analysis of retrieval of conditioned taste aversion in rats. Unit activity changes in critical brain regions. *Physiologia Bohemoslovaca, 28,* 525–536.

Buzsaki, G., Gage, F. H., Czopf, J., and Björklund, A. (1987). Restoration of rhythmic slow activity in the subcortically denervated hippocampus by fetal CNS transplants. *Brain Research, 400,* 334–347.

Cabanac, M. (1979). Physiological role of pleasure. *Science, 173,* 1103–1107.

Cahill, L. and McGaugh, J. L. (1990). Amygdaloid complex lesions differentially affect retention of tasks using appetitive and aversive reinforcement. *Behavioral Neuroscience, 104,* 532–543.

Cannon, D. S., Baker, T. B., and Berman, R. F. (1977). Taste aversion disruption by drug pretreatment: dissociative and drug-specific effects. *Pharmacology, Biochemistry and Behavior, 6,* 93–100.

Cappell, H. and LeBlanc, A. E. (1971). Conditioned aversion to saccharin by single administrations of mescaline and D-amphetamine. *Psychopharmacologia, 22,* 352–356.

Cappell, H. and LeBlanc, A. E. (1973). Punishment of saccharin drinking by amphetamine in rats and its reversal by chlordiazepoxide. *Journal of Comparative and Physiological Psychology, 85,* 97–104.

Cappell, H., LeBlanc, A. E., and Endrenyi, L. (1973). Aversive conditioning by psychoactive drugs: effects of morphine, alcohol and chlordiazepoxide. *Psychopharmacologia, 29,* 239–246.

Capretta, P. J. (1977). Establishment of food preferences by exposure to ingestive stimuli early in life. In: L. M. Barker, M. R. Best, and M. Domjan (Eds.), *Learning mechanisms in food selection* (pp. 99–121). Waco: Baylor University Press.

Carey, R. J. and Goodall, E. B. (1974). Amphetamine-induced taste aversion: a comparison of D- versus L-amphetamine. *Pharmacology, Biochemistry and Behavior, 2,* 325–330.

Castro, A. J., Zimmer, J., Sunde, N. A., and Bold, E. L. (1985). Transplantation of fetal cortex to the brain of newborn rats: a retrograde fluorescent analysis of callosal and thalamic projections from transplant to host. *Neuroscience Letters, 60,* 283–288.

Castro, A. J., Tonder, N., Sunde, N. A., and Zimmer, J. (1988). Fetal neocortical transplants grafted to the cerebral cortex of newborn rats receive afferents from the basal forebrain, locus coeruleus and midline raphe. *Experimental Brain Research, 69,* 613–622.

Castro, A. J., Sorensen, J. C., Tonder, N., Bold, L., and Zimmer, J. (1989). Fetal neocortical transplants grafted into cortical lesion cavities made in newborn rats receive multiple host afferents. A retrograde fluorescent tracer analysis. *Restorative Neurology and Neuroscience, 1,* 13–23.

Castro, A. J., Hogan, T. P., Sorensen, J. Chr., Klausen, B. S., Danielsen, E. H., Zimmer, J., and Neafsey, E. J. (1991). Heterotopic neocortical transplants. An anatomical and electrophysiological analysis of host projection to occipital cortical grafts placed into sensoriomotor cortical lesions made in newborn rats. *Developmental Brain Research, 58,* 231–236.

Caza, P., Brown, L., and Spear, N. E. (1982). Epinephrine-induced conditioned taste aversion. *Hormones and Behavior, 16,* 31–45.

Cechetto, D. F. (1987). Central representation of visceral function. *Federation Proceedings, 46,* 17–23.

Cechetto, D. F. and Saper, C. B. (1987). Evidence for a viscerotopic sensory representation in the cortex and thalamus in the rat. *Journal of Comparative Neurology, 262,* 27–45.

Cechetto, D. F., Standaert, D. G., and Saper, C. B. (1985). Spinal and trigeminal dorsal horn projections to the parabrachial nucleus in the rat. *Journal of Comparative Neurology, 240,* 153–160.

Chambers, K. C. (1990). A neural model for conditioned taste aversions. *Annual Review of Neuroscience, 13,* 373–385.

Chambert, G., Kobashi, M., and Adachi, A. (1993). Convergence of gastric and hepatic information in brain stem neurons of the rat. *Brain Research Bulletin, 32,* 525–529.

Chang, F.-C. T. and Scott, T. R. (1984). Conditioned taste aversions modify neural responses in the rat nucleus tractus solitarius. *Journal of Neuroscience, 4,* 1850–1862.

Chang, F. F., Steedman, J. G., and Lund, R. D. (1984). Embryonic cerebral cortex placed in the occipital region of newborn rats makes connections with the host brain. *Developmental Brain Research, 13,* 164–166.

Chang, F. F., Steedman, J. G., and Lund, R. D. (1986). The lamination and connectivity of embryonic cerebral cortex transplanted into newborn rat cortex. *Journal of Comparative Neurolology, 244,* 401–411.

Cherkin, A. (1969). Kinetics of memory consolidation: role of amnesic treatment parameters. *Proceedings of the National Academy of Sciences USA, 63,* 1094–1100.

Chiel, H. J. and Susswein, A. J. (1993). Learning that food is inedible in freely behaving Aplysia californica. *Behavioral Neuroscience, 107,* 327–338.

Chipkin, R. E. and Rosecrans, J. A. (1978). Aversiveness of oral methadone in rats. *Psychopharmacology, 57,* 303–310.

Clody, D. E. and Vogel, J. R. (1973). Drug-induced conditioned aversion to mouse-killing in rats. *Pharmacology, Biochemistry and Behavior, 1,* 477–481.

Coil, J. D., and Norgren, R. (1981). Taste aversions conditioned with intravenous copper sulfate: attenuation by ablation of the area postrema. *Brain Research, 212,* 425–433.

Coil, J. D., Rogers, R. C., García, J., and Novin, D. (1978). Conditioned taste aversions: vagal and circulatory mediation of the toxic unconditioned stimulus. *Behavioral Biology, 24,* 509–519.

Cole, A. J., Saffen, D. W., Baraban, J. M., and Worley, P. F. (1989). Rapid increase of an immediate early gene messenger RNA in hippocampal neurons by synaptic NMDA receptor activation. *Nature, 340,* 474–476.

Colley, P. and Routtenberg, A. (1993). Long-term potentiation as synaptic dialogue. *Brain Research Review, 18,* 115–122.

Connor, D. J., Langlais, P. J., and Thal, L. J. (1991). Behavioral impairments after lesions of the nucleus basalis by ibotenic acid and quisqualic acid. *Brain Research, 555,* 84–90.

Contreras, R. J., Beckstead, R. M., and Norgren, R. (1982). The central projections of the trigeminal, facial, glossopharyngeal and vagus nerves: an autoradiographic study in the rat. *Journal of Autonomic Nervous System, 6,* 303–322.

Corcoran, M. E. (1973). Role of drug novelty and metabolism in the aversive effects of hashish injections in rats. *Life Sciences, 12,* 63–72.

Corcoran, M. E., Bolotow, I., Amit, Z., and McCaughran, J. A., Jr. (1974). Conditioned taste aversions produced by active and inactive cannabinoids. *Pharmacology, Biochemistry and Behavior, 2,* 725–728.

Cross, R. J., Markesberry, W. R., Brooks, W. H., and Roszman, T. L. (1980). Hypothalamic immune interactions. *Brain Research, 196,* 79–87.

Cunningham, C. L. (1978). Alcohol interacts with flavor during extinction of conditioned taste aversion. *Physiological Psychology, 6,* 510–516.

Curtis, K. S., Sved, A. F., Verbalis, J. G., and Stricker, E. M. (1994). Lithium chloride-induced anorexia, but not conditioned taste aversions, in rats with area postrema lesions. *Brain Research, 663,* 30–37.

Dantzer, R. (1980). Conditioned taste aversion as an index of lead toxicity. *Pharmacology, Biochemistry and Behavior, 13,* 133–135.

Davis, M. (1994). The role of the amygdala in emotional learning. *International Review of Neurobiology, 36,* 225–266.

Davis, J. L. and Bures. J. (1972). Disruption of saccharin aversion learning in rats by cortical spreading depression in the CS–US interval. *Journal of Comparative and Physiological Psychology, 80,* 398–402.

Davis, H. P. and Squire, L. R. (1984). Protein synthesis and memory: a review. *Psychological Bulletin, 96,* 518–559. Decker, M. W. and McGaugh, J. L. (1991). The role of interactions between the cholinergic and other neuromodulatory systems in learning and memory. *Synapse, 7,* 151–168.

De La Casa, G. and Lubow, R. E. (1995). Latent inhibition of conditioned taste aversion: the roles of stimulus frequency and duration and the amount of fluid ingested during preexposure. *Neurobiology of Learning and Memory, 64,* 125–132.

Delaney, K. and Gelperin A. (1986). Post-ingestive food aversion learning in amino-deficiemt diets by the terrestrial slug *Limax maximus. Journal of Comparative Physiology A, 159,* 281–295.

De Luca, B. and Monda, M. (1986). Neophobia, conditioned taste aversion and EEG arousal after globus pallidus lesion. *Physiology and Behavior, 36,* 545–551.

de Montigny, C. (1989). Cholecystokinin tetrapeptide induce pain-like attacks in healthy volunteers. *Archives of Neurology and Psychiatry, 46,* 511–517.

Deutsch, J. A. and Hardy, W. T. (1977). Cholecystokinin produces bait shyness in rats. *Nature, 266,* 196.

Deutsch, J. A. and Parsons, S. L. (1981). Bombesin produces taste aversion in rats. *Behavioral and Neural Biology, 31,* 110–113.

Deutsch, J. A., Molina, F., and Puerto, A. (1976). Conditioned taste aversion caused by palatable nontoxic nutrients. *Behavioral Biology, 16,* 161–174.

De Wied, D. (1969). Effects of peptide hormones on behavior. In W. F. Ganog and L. Martini (Eds.), *Frontiers in neuroendocrinology* (pp. 97–180). Oxford: Oxford University Press.

Dickinson, A. (1988). Intentionality in animal conditioning. In L. Weiskrantz (Ed.) *Thought without language.* Oxford: Oxford University Press.

Di Lorenzo, P. M. (1985). Responses to NaCl of parabrachial units that were conditioned with intravenous LiCl. *Chemical Senses, 10,* 438.

Di Lorenzo, P. M. (1988*a*). Long-delay learning in rats with parabrachial pontine lesions. *Chemical Senses, 13,* 219–229.

Di Lorenzo, P. M. (1988*b*). Taste responses in the parabrachial pons of decerebrate rats. *Journal of Neurophysiology, 59,* 1871–1887.

Di Lorenzo, P. M. (1990). Corticofugal influence on taste responses in the parabrachial pons of the rat. *Brain Research, 530,* 73–84.

Di Lorenzo, P. M., and Monroe, S. (1992). Corticofugal input to taste responsive units in the parabrachial pons. *Brain Research Bulletin, 29,* 925–930.

Divac, I., Gade, A., and Wikmark, R. E. G. (1975). Taste aversion in rats with lesions in the frontal lobes: no evidence for interoceptive agnosia. *Physiological Psychology, 3,* 43–46.

D'Mello, G. D., Goldberg, D. M., Goldberg, S. R., and Stolerman, I. P. (1979). Conditioned taste aversion and operant behaviour in rats: effects of cocaine and a cocaine analogue (Win 35, 428). *Neuropharmacology, 18,* 1009–1010.

D'Mello, G. D., Goldberg, D. M., Goldberg, S. R., and Stolerman, I. P. (1981). Conditioned taste aversion and operant behavior in rats: effects of cocaine, apomorphine and some long acting derivatives. *Journal of Pharmacology and Experimental Therapeutics, 219,* 60–68.

Domjan, M. (1972). CS prexposure in taste aversion learning: effects of deprivation and preexposure duration. *Learning and Motivation, 3,* 389–402.

Domjan, M. (1977). Attenuation and enhancement of neophobia for edible substances. In L. Barker, M. Best, and M. Domjan (Eds.), *Learning mechanisms in food selection* (pp. 151–179). Waco: Baylor University Press.

Domjan, M. (1980). Effects of the intertrial interval on taste-aversion learning in rats. *Physiology and Behavior, 25,* 117–125.

Domjan, M. and Levy, C.J. (1977). Taste aversions conditioned by the aversiveness of insulin and formalin: role of CS specificity. *Journal of Experimental Psychology: Animal Behavior Processes, 3,* 119–131.

Domjan, M. and Wilson, N. E. (1972). Contribution of ingestive behaviors to taste-aversion learning. *Journal of Comparative and Physiological Psychology, 80,* 403–412.

Doty, R. W. and Ringo, J. L. (1994). Hemispheric distribution of memory traces. In J. Delacour (Ed.), *The memory systems of the brain* (pp. 636–656). Singapore: World Scientific.

Dragoin, W., McCleary, G. E., and McCleary, P. A. (1971). A comparison of two methods of measuring conditioned taste aversions. *Behavior Research Methods and Instrumentation, 3,* 309–310.

Dudai, Y. (1989). *The neurobiology of memory. Concepts, findings, trends.* Oxford: Oxford University Press.

Dudai, Y. (1996). Consolidation: fragility on the road to the engram. *Neuron, 17,* 367–370.

Dunn, L. T. and Everitt, B. J. (1988). Double dissociations of the effects of amygdala and insular cortex lesions on conditioned taste aversion, passive avoidance and neophobia in the rat using excitotoxin ibotenic acid. *Behavioral Neuroscience, 102,* 3–23.

Dunnett, S. B. and Björklund, A. (1987). Mechanisms of function of neural grafts in the adult mammalian brain. *Journal of Experimental Biology, 132,* 256–289.

Dunnett, S. B. and Björklund, A. (1994). *Functional neural transplantation.* New York: Raven Press,

Dunnett, S. B. and Richards, S. J. (Eds.) (1990). Neural transplantation: from molecular basis to clinical applications. *Progress in brain research, Vol. 82.* Oxford: Elsevier.

Durkin, T. P. and Toumane, A. (1992). Septo-hippocampal and nBM cortical cholinergic neurones exhibit differential time-courses of activation as a function of both type and duration of spatial memory testing in mice. *Behavioural Brain Research, 50,* 43–52.

Elkins, R. L. (1973). Attenuation of drug induced bait shyness to a palatable solution as an increasing function of its availability prior to conditioning. *Behavioral Biology, 9,* 221–226.

Elkins, R. L. (1980). Attenuation of X-ray induced taste aversions by the olfactory bulb or amygdaloid lesion. *Physiology and Behavior, 24,* 515–521.

Elkins, R. L., Walters, P. A., and Ott, T. E. (1992). Continued development and unconditioned stimulus characterization of selectively bred lines of taste aversion prone and resistant rats. *Alcohol: Clinical and Experimental Research, 16,* 928–934.

Emmerick, J. J. and Snowdon, C. T. (1976). Failure to show modification of male golden hamster mating behavior through taste–odor aversion learning. *Journal of Comparative and Physiological Psychology, 90,* 857–869.

Erickson, R. P. (1963). Sensory neural patterns and gustation. In Y. Zotterman (Ed.), *Olfaction and taste, I* (pp. 205–213). Oxford: Pergamon Press.

Ervin, G. N., Mosher, J. T., Birkemo, L. S., and Jonhson, F. M. (1995). Multiple small doses of cholecystokinin octapeptide are more efficacious at inducing taste aversion conditioning than single large doses. *Peptides, 16,* 539–545.

Escobar, M. L., Fernández, J., Guevara-Aguilar, R., and Bermúdez-Rattoni, F. (1989). Fetal brain grafts induce recovery of learning deficits and connectivity in rats with gustatory neocortex lesion. *Brain Research, 478,* 368–374.

Escobar, M. L., Jiménez, N., López-García, J. C., Tapia, R., and Bermúdez-Rattoni, F. (1993). Nerve growth factor with insular cortical grafts induces recovery of learning and reestablishes graft choline acetyltransferase activity. *Journal of Neural Transplantation and Plasticity, 2,* 167–172.

Escobar, M. L., Russell, R. W., Booth, R. A., and Bermúdez-Rattoni, F. (1994). Accelerating behavioral recovery after cortical lesions. I. Homotopic implants

plus NGF. *Behavioral and Neural Biology, 61,* 73–80.
Etscorn, F. (1980). Sucrose aversions in mice as a result of injected nicotine or passive tobacco smoke inhalation. *Bulletin of the Psychonomic Society, 15,* 54–56.
Etscorn, F. and Parson, P. (1979). Taste aversion in mice using phencyclidine and ketamine as the aversive agents. *Bulletin of the Psychonomic Society, 14,* 19–21.
Ettenberg, A., van der Kooy, D., Le Moal, M., Koob, G., and Bloom, F. (1983). Can aversive properties of (peripherally-injected) vasopression account for its putative role in memory? *Behavioural Brain Research, 7,* 331–350.
Evenden, J. L., Lavis, L., and Iversen, S. D. (1992). Blockade of conditioned taste aversion by scopolamine and *N*-methylscopolamine: associative conditioning, not amnesia. *Psychopharmacology, 106,* 179–188.
Feinberg, A. (1973). Effect of treatment-test interval and proferrin on X-irradiation-induced saccharin aversion. *JSAS Catalog of Selected Documents in Psychology, 3,* 119–141.
Felten, D. L., Ackerman, K. D., Wiegand, S. J., and Felten, S. Y. (1987). Noradrenergic and sympathetic innervation of the spleen. *Neuroscience Research, 18,* 28–36.
Fernández-Ruíz, J., Escobar, M. L., Piña, A. L., Díaz Cintra-McGlone, F. L., and Bermúdez-Rattoni, F. (1991). Time-dependent recovery of taste aversion learning by fetal brain transplants in gustatory neocortex. *Behavioral and Neural Biology, 55,* 179–193.
Fernández-Ruíz, J., Miranda, M. I., Bermúdez-Rattoni, F., and Drucker-Colín R. (1993). Effects of catecholaminergic depletion of the amygdala and insular cortex on the potentiation of odor by taste aversions. *Behavioral and Neural Biology, 60,* 189–191.
Fernández-Ruíz, J., Guzmán, R., Martínez, M. D., Miranda, M. I., Bermúdez-Rattoni, F. and Drucker-Colín, R. (1994). Adrenal medullary grafts restore olfactory deficits and catecholamine levels of 6-OHDA amygdala lesioned animals. *Journal of Neural Transplantation and Plasticity, 4,* 289–297.
Ferry, B., Sandner, G., and DiScala, G. (1995). Neuroanatomical and functional specificity of the basolateral amygdaloid nucleus in taste-potentiated odor aversion. *Neurobiology of Learning and Memory, 64,* 169–180.
Fishberg, A. M., Hitzig, W. M., and King, R. H. (1933). Measurement of the circulation time with saccharin. *Proceedings of the Society for Experimental Biology and Medicine, 30,* 651–652.
Fitzgerald, R. E. and Burton, M. J. (1983). Neophobia and conditioned taste aversion deficits in the rat produced by undercutting temporal cortex. *Physiology and Behavior, 30,* 203–206.
Flaherty, C. F., Turovsky, J., and Kraus, K. L. (1994). Relative hedonic value modulates anticipatory contrast. *Physiology and Behavior, 55,* 1047–1054.
Floeter, M. and Jones, E. G. (1984). Connections made by transplants to the cerebral cortex of rat brains damaged *in utero. Journal of Neuroscience, 4,* 141–150.
Floeter, M. and Jones, E. G. (1985). Transplantation of fetal postmitotic neurons

to rat cortex: survival early pathway choices and long term projections of outgrowing axons. *Developmental Brain Research, 22,* 19–38.

Flynn, F. W., Grill, H. J., Schulkin, J., and Norgren, R. (1991). Central gustatory lesions. II. Effects on sodium appetite, taste aversion learning, and feeding behaviors. *Behavioral Neuroscience, 105,* 944–954.

Foltin, R. W. and Schuster, C. R. (1981). The effects of d1-cathinone in a gustatory avoidance paradigm. *Pharmacology, Biochemistry and Behavior, 14,* 907–909.

Foltin, R. W. and Schuster, C. R. (1982). The effects of cocaine in a gustatory avoidance paradigm: a procedural analysis. *Pharmacology, Biochemistry and Behavior, 16,* 347–352.

Forkman, B. (1993). The effect of uncertainty on the food intake of the mongolian gerbil. *Behaviour, 124,* 197–207.

Forkman, B. (1995). The effect of hunger on the learning of new food preferences in the mongolian gerbil. *Behaviour, 132,* 627–639.

Fox, R. A., Corcoran, M., and Brizzee, K. (1990). Conditioned taste aversion and motion sickness in cats and squirrel monkeys. *Canadian Journal of Physiology and Pharmacology, 68,* 269–278.

Frank, M. (1973). An analysis of hamster afferent taste nerve response functions. *Journal of General Physiology, 61,* 588–618.

Frank, M. E. and Nowlis, G. H. (1989). Learned aversions and taste qualities in hamsters. *Chemical Senses, 14,* 379–394.

Freedman, N. L. and Ward, R. (1976). Saccharin aversion produced bys reversible functional cortical ablation. *Physiological Psychology, 4,* 485–488.

Frenk, H. and Rogers, G. H. (1979). The suppressant effects of naloxone on food and water intake in the rat. *Behavioral and Neural Biology, 26,* 23–40.

Frumkin, K. (1975). Failure of sodium- and calcium-deficient rats to acquire conditioned taste aversions to the object of their specific hunger. *Journal of Comparative and Physiological Psychology, 89,* 329–339.

Gage, F. H. (1990). NGF-dependent sprouting and regeneration in the hippocampus. *Progress in Brain Research, 83,* 357–370.

Gage, F. H. and Björklund, A. (1986). Enhanced graft survival in the hippocampus following selective denervation. *Neurosciences, 17,* 89–98.

Galaverna, O. G., Seeley, R. J., Berridge, K. C., Grill, H. J., Epstein, A. N., and Schulkin, J. (1993). Lesions of the central nucleus of the amygdala. I. Effects on taste reactivity, taste aversion learning and sodium appetite. *Behavioral Brain Research, 59,* 11–17.

Gallo, M. and Bures, J. (1991). Acquisition of conditioned taste aversion in rats is mediated by ipsilateral interaction of cortical and mesencephalic mechanisms. *Neuroscience Letters, 133,* 187–190.

Gallo, M., and Candido, A. (1995*a*). Reversible inactivation of dorsal hippocampus by tetrodotoxin impairs blocking of taste aversion selectively during the acquisition but not the retrieval in rats. *Neuroscience Letters, 186,* 1–4.

Gallo, M., and Candido, A. (1995*b*). Dorsal hippocampal lesions impair blocking but not latent inhibition of taste aversion learning in rats. *Behavioral Neuroscience, 109,* 413–425.

Gallo, M., Arnedo, M., Aguero, A., and Puerto, A. (1988). Electrical intracerebral stimulation of the area postrema on taste aversion learning. *Behavioural Brain Research, 30,* 289–296.

Gallo, M., Arnedo, M., Aguero A., and Puerto A. (1990). The functional relevance of the area postrema in drug-induced aversion learning. *Pharmacology Biochemistry and Behavior, 35,* 543–551.

Gallo, M., Arnedo, M., Aguero A., and Puerto A. (1991). Participation of the area postrema in learned aversions induced by body rotation. *Behavioural Brain Research, 42,* 13–23.

Gallo, M., Roldan, G., and Bures, J. (1992). Differential involvement of gustatory cortex and amygdala in acquisition and retrieval of conditioned taste aversion in rats. *Behavioural Brain Research, 52,* 91–97.

Gamzu, E. (1977). The multifaceted nature of taste-aversion inducing agents: is there a single common factor? In L. Barker, M. Best, and M. Domjan (Eds.), *Learning mechanisms in food selection* (pp. 477–509). Waco, Texas: Baylor University Press.

Gamzu, E., Vincent, G., and Boff, E. (1985). A pharmacological perspective of drugs used in establishing conditioned food aversions. *Annals of the New York Academy of Sciences, 443,* 231–249.

Ganchrow, D. and Erickson, R. P. (1972). Thalamocortical relations in gustation. *Brain Research, 36,* 289–305.

García, J. (1989). Food for Tolman: cognition and cathesis in concert. In T. Archer and L. G. Nilson (Eds.), *Aversion, avoidance and anxiety* (pp. 45–85). Hillsdale, New Jersey: Lawrence Erlbaum Associates.

García, J. (1990). Learning without memory. *Journal of Cognitive Neuroscience, 2,* 287–304.

García, J. and Ervin, F. R. (1968). Gustatory-visceral and telereceptor-cutaneous conditioning: adaptation in the internal and external milieus. *Communications in Behavioral Biology, 1,* 389–415.

García, J. and Kimeldorf, D. J. (1960). Conditioned avoidance behaviour induced by low-dose neutron exposure. *Nature, 185,* 261–262.

García, J. and Koelling, R. A. (1966). Relation of cue to consequence in avoiding learning. *Psychonomic Science, 4,* 123–124.

García, J., Kimmeldorf, D. J., and Koelling, R. A. (1955). Conditioned aversion to saccharin resulting from exposure to gamma radiation. *Science, 122,* 157–158.

García, J., Ervin, F. R., and Koelling, R. A. (1966). Learning with prolonged delay of reinforcement. *Psychonomic Science, 5,* 121–122.

García, J., Ervin, F. R., and Koelling, R. A. (1967). Toxicity of serum from irradiated donors. *Nature, 213,* 682–683.

García, J., Hankins, W. G., and Rusiniak, K. W. (1974). Behavioral regulation of the milieu interne in man and rat. *Science, 185,* 824–831.

García, J., Rusiniak, K. W., Kiefer, S. W., and Bermúdez-Rattoni, F. (1982). The neural integration of feeding and drinking habits. In C. Woody (Ed.), *Conditioning* (pp. 567–579). New York: Plenum Press.

García, J., Lasiter, P. S., Bermúdez-Rattoni, F., and Deems, D. A. (1985). A general theory of aversions learning. In M. S. Braveman and P. Bronstein (Eds.), Experimental assessments and clinical applications of conditioned food aversions. *Annals of the New York Academy of Sciences, 443,* 8–21.

García, J., Brett, L. P., and Rusiniak, K. (1989) Limits of Darwinian conditioning. In S.B. Klein and R. Moners (Eds.), *Contemporary learning theory* (pp. 181–203). Hillsdale, New Jersey: Lawrence Erlbaum.

Garraud, P., Gray, J. A., and De Wied, D. (1974). Pituitary adrenal hormones and extinction of rewarded behavior in the rat. *Physiology and Behavior,* 12, 109–119.

Gaston, K.E. (1978). Brain mechanisms of conditioned taste aversion learning: a review of the literature. *Physiological Psychology, 6,* 340–353.

Gay, P. E., Leaf, R. C., and Arble, F. B. (1975). Inhibitory effects of pre- and posttest drugs on mouse killing by rats. *Pharmacology, Biochemistry and Behavior, 3,* 33–45.

Gelperin, A. (1975). Rapid food-aversion learning by a terrestrial mollusk. *Science, 189,* 567–520.

Gold, R. and Proulx, D. (1972). Bait-shyness acquisition is impaired by VMH lesions that produce obesity. *Journal of Comparative and Physiological Psychology, 79,* 201–209.

Golus, P. and McGee, R. (1980). Metrazol produces conditioned taste aversion in rats. *Psychopharmacology, 68,* 257–259.

Goodard, G. (1969). Analysis of avoidance conditioning following cholinergic stimulation of amygdala in rats. *Journal of Comparative and Physiological Psychology, 69,* 1–18.

Gosnell, B. A. and Hsiao, S. (1984). The effects of cholecystokinin on taste preference and sensitivity in rats. *Behavioral Neuroscience, 3,* 452–460.

Goudie, A. J. (1981). Stimulus properties of cocaine/chlordiazepoxide mixtures in rodents. *IRCS Medical Sciences: Biochemistry; Drug Metabolism and Toxicology; Nervous System; Pharmacology, Psychology and Psychiatry, 9,* 663–664.

Goudie, A. J., Thornton, E. W., and Wheeler, T. J. (1976). Drug pretreatment effects in drug induced taste aversions: effects of dose and duration of pretreatment. *Pharmacology, Biochemistry and Behavior, 4,* 629–633.

Grant, V. L. (1987). Do conditioned taste aversions result from activation of emetic mechanisms? *Psychopharmacology, 93,* 405–415.

Green, K. F. and Parker, L. A. (1975). Gustatory memory: incubation and interference. *Behavioral Biology, 13,* 359–367.

Greenshaw, A. J., Aleksanyan, Z. A., Kundu, S. N., Bracha, V., and Bures, J. (1985). A response specific conditioned aversion to hypothalamic stimulation in rats. *Brain Research, 339,* 130–135.

Grigson, P. S. (1997) Conditioned taste aversions and drugs of abuse: a reinterpretation. *Behavioral Neuroscience, 111,* 129–136.

Grigson, P. S., Shimura, T., and Norgren, R. (1997*a*). Brainstem lesions and gustatory function. II. The role of the nucleus of the solitary tract in Na^+ appetite, conditioned taste aversion and conditioned odor aversion in rats. *Behavioral Neuroscience, 111,* 169–179.

Grigson, P. S., Shimura, T., and Norgren, R. (1997*b*). Brainstem lesions and gustatory function: III. The role of the nucleus of the solitary tract and the parabrachial nucleus in retention of a conditioned taste aversion in rats. *Behavioral Neuroscience, 111,* 180–187.

Grill, H. J. (1985). Discussion of physiological mechanisms in conditioned taste aversions. *Annals of the New York Academy of Sciences, 443,* 67–88.

Grill, H. J. and Berridge, K. C. (1981). Chronic decerebrate rats demonstrate preabsorptive insulin secretion and hyperinsulinemia. *Society for Neuroscience Abstracts, 7,* 29.

Grill, H. J. and Berridge, K. C. (1985). Taste reactivity as a measure of the neural control of palatability. *Progress in Psychobiology and Physiological Psychology, 11,* 1–61.

Grill, H. J. and Norgren, R. (1978*a*). The taste reactivity test. I. Mimetic responses to gustatory stimuli in neurologically normal rats. *Brain Research, 143,* 263–279.

Grill, H. J. and Norgren, R. (1978*b*). The taste reactivity test. II. Mimetic responses to gustatory stimuli in chronic thalamic and chronic decerebrate rats. *Brain Research, 143,* 281–297.

Grill, H. J. and Norgren, R. (1978*c*). Chronically decerebrate rats demonstrate satiation but not bait shyness. *Science, 201,* 267–269.

Groenewegen, H. J. (1988). Organization of the afferent connections of the mediodorsal thalamic nucleus in the rat, related to the mediodorsal-prefrontal topography. *Neuroscience, 24,* 379–431.

Groenewegen, H. J. and Berendse, H. W. (1994). The specificity of the 'nonspecific' midline and intralaminar thalamic nuclei. *Trends in Neurosciences, 17,* 52–57.

Groenewegen, H. J., Berendse, H. W., Wolters, J. G., and Lohman, A. H. M. (1990). The anatomical relationship of the prefrontal cortex with the striatopallidal system, the thalamus and the amygdala: evidence for a parallel organization. *Progress in Brain Research, 85,* 95–118.

Grossman, S. P. (1964). Behavioral effects of chemical stimulation of ventral amygdala. *Journal of Comparative and Physiological Psychology, 57,* 29–36.

Grote, F. W., Jr. and Brown, R. T. (1971). Conditioned taste aversions: two-stimulus tests are more sensitive than one-stimulus tests. *Behavior Research Methods and Instrumentation, 3,* 311–312.

Grote, F. W., Jr. and Brown, R. T. (1973). Deprivation level affects extinction of a conditioned taste aversion. *Learning and Motivation, 4,* 314–319.

Grupp, L. A., Linseman, M. A., and Cappell, H. (1976). Effects of amygdala lesions on taste aversions produced by amphetamine and LiCl. *Pharmacology, Biochemistry and Behavior, 4,* 541–544.

Gu, Y., Gonzalez, F. G., Chen, D. Y., and Deutsch, J. A. (1993). Expression of c-*fos* in brain subcortical structures in response to nauseant lithium chloride and osmotic pressure in rats. *Neuroscience Letters, 157,* 49–52.

Gustavson, C. R., Jowsey, J. R., and Milligan, D. N. (1982). A 3-year evaluation of taste aversion coyote control in Saskatchewan. *Journal of Range Management, 35,* 57–59.

Gustavson, C. R., Gustavson, J. C., and Holzer, G. A. (1983). Thiabandazole-based taste aversions in dingoes (*Canis familiaris dingo*) and New Guinea wild dogs (*Canis familiaris hallstromi*). *Applied Animals Ethology, 10,* 385–388.

Gustavson, C. R., Gustavson, J. C., Young, J. K., Pumariega, A. J., and Nicolaus, I. K. (1989). Estrogen induce malaise. In J. M. Lakoski, J. R. Perez-Polo, D. K. Rassin, C. R. Gustavson, and C. S. Watson (Eds.), *Neural control of reproductive function* (pp. 501–523). New York: Alan R. Liss.

Gustavson, K. K., Hart, J. A., Calton, J. L., and Schachtman, T. R. (1992). Effects of extinction and US reinstatement of a blocking CS–US association. *Bulletin of the Psychonomic Science, 30,* 247–250.

Haberly, L. B. and Price, J. L. (1978). Association and commissural fiber systems of the olfactory cortex in the rat. II. Systems originating in the olfactory peduncle. *Journal of Comparative Neurology, 181,* 781–808.

Hagg, T., Vahlsing, H. L., Manthorpe, M., and Varon, S. (1990). Nerve growth factor infusion into denervated adult rat hippocampal formation promotes its cholinergic reinervation. *Journal of Neuroscience, 10,* 3087–3092.

Halpern, B. P. and Tapper, D. N. (1971). Taste stimuli: quality coding time. *Science, 171,* 1256–1258.

Hamilton, R. B. and Norgren, R. (1984). Central projections of gustatory nerves in the rat. *Journal of Comparative Neurology, 222,* 560–577.

Hankins, W. G., García, J., and Rusiniak, K. W. (1974). Cortical lesions: flavor–illness and noise–shock conditioning. *Behavioral Biology, 10,* 173–181.

Harrer, M. I. and Travers, S. P. (1996). Topographic organization of fos-like immunoreactivity in the rostral nucleus of the solitary tract evoked by gustatory stimulation with sucrose and quinine. *Brain Research, 711,* 125–137.

Harrison, T. A. and Miller, N. W. (1994). Fos protein expression in taste neurons following electrical stimulation of afferent nerves. In K. Kurihara, N. Suzuki, and H. Ogawa (Eds.), *Olfaction and taste, XI* (pp. 410–411). Tokyo: Springer-Verlag.

Hatfield, T. and Gallagher, M. (1995). Taste potentiated odor conditioning: impairment produced by infusion of an *N*-methyl-D-aspartate antagonist into basolateral amygdala. *Behavioral Neuroscience, 109,* 663–668.

Hatfield, T., Graham, P. W., and Gallagher, M. (1992). Taste potentiated odor aversion learning: role of amygdala basolateral complex and central nucleus. *Behavioral Neuroscience, 106,* 286–293.

Hellekant, G. and Ninomiya, Y. (1991). On the taste of umami in chimpanzee. *Physiology and Behavior, 49,* 927–934.

Helmstetter, F. J. and Bellgowan, P. S. (1994). Effects of muscimol applied to the basolateral amygdala on acquisition and expression of contextual fear conditioning in rats. *Behavioral Neuroscience, 108,* 1005–1009.

Herbert, H., Moga, M. M., and Saper, C. B. (1990). Connections of the parabrachial nucleus with the nucleus of the solitary tract and the medullary reticular formation in the rat. *Journal of Comparative Neurology, 293,* 540–580.

Hermann, G. E. and Rogers, R. C. (1985). Convergence of vagal and gustatory afferent input within the parabrachial nucleus of the rat. *Journal of Autonomic Nervous System, 13,* 1–17.

Hermann, G. E., Kohlerman, N. J., and Rogers, R. C. (1983). Hepatic–vagal and gustatory afferent interactions in the brainstem of the rat. *Journal of Autonomic Nervous System, 9,* 477–495.

Hernadi, I., Karadi, Z., Faludi, B., Vigh, J., Fekete, E., Galosi, R., Fogarasy, A., and Lenard, L. (1995). Deficits of conditioned taste aversion after prefrontal microlesions in the rat. *European Journal of Neuroscience, S8,* 94.

Hernadi, I., Karadi, Z., Faludi, B., and Lenard, L. (1997). Disturbances of neophobia and taste aversion learning after bilateral kainate microlesions in the rat pallidum. *Behavioral Neuroscience, 111,* 137–146.

Holder, M. D. Bermúdez-Rattoni, F., and García, J. (1988). Taste-potentiated noise–illness associations. *Behavioral Neuroscience, 102,* 363–371.

Houpt, T. A., Philopena, J. M., Wessel, T. C., Joh, T. H., and Smith, G. P. (1994). Increased c-*fos* expression in nucleus of the solitary tract correlated with conditioned taste aversion to sucrose in rats. *Neuroscience Letters, 172,* 1–5.

Houpt, T. A., Berlin, R. A., and Smith, G. P. (1995). c-*fos* induction in the nucleus of the solitary tract by sucrose after conditioned taste aversion acquisition is not attenuated by subdiaphragmatic vagotomy. *Chemical Senses, 20,* 710–711.

Houpt, T. A., Philopena, J. M., Wessel, T. C., Joh, T. H., and Smith, G. P. (1996). c-*fos* induction in the rat nucleus of the solitary tract by intraoral quinine infusion depends on prior contingent pairing of quinine and lithium chloride. *Physiology and Behavior, 60,* 1535–1541.

Howard, W. E., Marsh, R. E., and Cole, R. E. (1977). Duration of associative memory to toxic bait in deer mice. *Journal of Wildlife Management, 41,* 484–486.

Howard, W. E., Palmateer, S. D., and Nachman, M. (1968). Aversion to strychnine sulfate by Norway rats, roof rats, and pocket gophers. *Toxicology and Applied Pharmacology, 12,* 229–241.

Hsiao, S. and Deupree, D. (1983). Cholecystokinin and bombesin effects on rewarded and nonrewarded operants. *Peptides, 4,* 1–3.

Hunt, T. and Amit, Z. (1987). Conditioned taste aversion induced by self-administered drugs: paradox revisited. *Neuroscience and Biobehavioral Reviews, 11,* 107–130.

Hunt, W. A., Rabin, B. M., and Lee, J. (1987). Ethanol-induced taste aversions: lack of involvement of acetaldehyde and the area postrema. *Alcohol, 4,* 169–173.

Hylden, J. L., Anton, F., and Nahin, R. L. (1989). Spinal lamina I projection neurons in the rat: collateral innervation of parabrachial area and thalamus. *Neuroscience, 28,* 27–37.

Ikeda, K. (1909). On a new seasoning. *Journal of Chemical Society of Japan, 27,* 279–284 (in Japanese).

Ionescu, E. and Buresova, O. (1977*a*). Failure to elicit conditioned taste aversion by severe poisoning. *Pharmacology, Biochemistry and Behavior, 6,* 251–254.

Ionescu, E. and Buresova, O. (1977*b*). Effects of hypothermia on the acquisition of conditioned taste aversion in rats. *Journal of Comparative and Physiological Psychology, 91,* 1297–1307.

Islam, S. (1978). Eliciting conditioned taste aversion by cobra venom neurotoxin in rats. *Indian Journal of Physiology and Pharmacology, 22,* 368–371.

Islam, S. (1980). Snake neurotoxins and conditioned taste aversion in mice. *International Journal of Neuroscience, 11,* 41–43.

Ito, S. (1992). Multiple projection of vagal non-myelinated afferents to the anterior insular cortex in rats. *Neuroscience Letters, 148,* 151–154.

Ivanova, S. F. and Bures, J. (1990*a*). Acquisition of conditioned taste aversion in rats is prevented by tetrodotoxin blockade of a small midbrain region centered around the parabrachial nuclei. *Physiology and Behavior, 48,* 543–549.

Ivanova, S. F. and Bures, J. (1990*b*). Conditioned taste aversion is disrupted by prolonged retrograde effects of intracerebral injection of tetrodotoxin in rats. *Behavioral Neuroscience, 104,* 948–954.

Izquierdo, I. (1993). Long-term potentiation and the mechanisms of memory. *Drug Development Research, 30,* 1–17.

Izquierdo, I. (1994). Pharmacological evidence for a role of long-term potentiation in memory. *The FASEB Journal, 8,* 1139–1145.

Izquierdo, I., Fin, C., Schmitz, P. K., Da Silva, R. C., Jerusalinsky, D., Quillfeldt, J. A., Ferreira, M. B. G., Medina, J. H., and Bazan, N. G. (1995). Memory enhancement by intrahippocampal, intraamygdala, or intraentorhinal infusion of platelet-activating factor measured in an inhibitory avoidance task. *Proceedings of the National Academy of Sciences USA, 92,* 5047–5051.

Jackson, A. and Sanger, D. J. (1989). Conditioned taste aversions induced by phencyclidine and other antagonists of *N*-methyl-D-aspartate. *Neuropharmacology, 28,* 459–464.

Jaeger, C.B. and Lund, R.D. (1981). Transplantation of embryonic occipital cortex to the brain of newborn rats: a Golgi study of mature and developing transplants. *Journal of Comparative Neurology, 200,* 231–230.

Jaeger, T. V. and van der Kooy, D. (1993). Morphine acts in the parabrachial nucleus, a pontine viscerosensory relay, to produce discriminative stimulus effects. *Psychopharmacology, 110,* 76–84.

Johnson, V. B. (1923). Further contribution to the study of the evolution of the forebrain. *Journal of Comparative Neurology, 35,* 337–481.

Johnson, G. V. W. and Jope, R. S. (1992). The role of microtubule-associated protein 2 (MAP-2) in neuronal growth, plasticity, and degeneration. *Journal of Neuroscience Research, 33,* 505–512.

Jones, E. G. (1975). Possible determinants of the degree of retrograde neuronal labeling with horseradish peroxidase. *Brain Research, 85,* 249–253.

Kalat, J. W. and Rozin, P. (1971). Role of interference in taste-aversion learning. *Journal of Comparative and Physiological Psychology, 77,* 53–58.

Kalia, M., and Sullivan, J. M. (1982). Brain stem projections of sensory and motor components of the vagus nerve in the rat. *Journal of Comparative Neurology, 211,* 248–264.

Kallman, M. J., Lynch, M. R., and Landauer, M. R. (1983). Taste aversions to several halogenated hydrocarbons. *Neurobehavioral Toxicology and Teratology, 5,* 23–28.

Kamin, L. J. (1969). Predictability, surprise, attention, and conditioning. In B. A. Campbell and R. M. Church (Eds.), *Punishment and aversive behavior* (pp. 279–296). New York: Appleton-Century-Crofts.

Kapp, B. S., Schwaber, J. S., and Driscoll, P. A. (1985). Frontal cortex projections to the amygdaloid central nucleus in the rabbit. *Neuroscience, 15,* 327–346.

Kapp, B.S., Whalen, P.J., Supple, W.F., and Pascoe, J.P. (1992). Amygdaloid contributions to conditioned arousal and sensory information processing. In J. P. Aggleton (Ed.), *The Amygdala: Neurobiological Aspects of Emotion, Memory and Mental Dysfunction,* (pp. 229–254). New York: Wiley.

Kawamura, Y. and Kare, M.R. (Eds.) (1987). *Umami: A Basic Taste* (pp. 1–649). New York: Marcel Dekker.

Kawamura, Y., Kurihara, K., Nicolaidis, S., Oomura, Y., and Wayner, M.J. (Eds.) (1991). Umami: Proceedings of the Second International Symposium on Umami. *Physiology and Behavior, 49,* Special Issue, 385–1030.

Kesner, R. P., Berman, R. F., Burton, B., and Hankins, W. G. (1975). Effects of electrical stimulation of amygdala upon neophobia and taste aversion. *Behavioral Biology, 13,* 349–358.

Kiefer, S. W. (1985). Neural mediation of conditioned food aversions. *Annals of the New York of Academy of Sciences, 443,* 100–109.

Kiefer, S. W. (1995). Alcohol, palatability, and taste reactivity. *Neuroscience and Behavioral Reviews, 19,* 133–141.

Kiefer, S. W. and Braun, J. J. (1979). Acquisition of taste avoidance habits in rats lacking gustatory neocortex. *Physiological Psychology, 7,* 245–250.

Kiefer, S. W. and Orr, M. R. (1992). Taste avoidance, but not aversion, learning in rats lacking gustatory cortex. *Behavioral Neuroscience, 106,* 140–146.

Kiefer, S. W., Rusiniak, K. W., and García, J. (1982). Flavor–illness aversions: potentiation of odor by taste in rats with gustatory neocortex ablations. *Journal of Comparative and Physiological Psychology, 96,* 540–548.

Kiefer, S. W., Leach, L. R., and Braun, J. J. (1984). Taste agnosia following gustatory neocortex ablation: dissociation from odor and generality across taste qualities. *Behavioral Neuroscience, 98,* 590–608.

Kiefer, S. W., Lawrence, G. J., and Metzler, C. W. (1986). Learned alcohol aversions in rats: gustatory and olfactory components. *Alcohol, 3,* 27–31.

Kiefer, S. W., Lawrence, G. J., and Metzler, C. W. (1987). Alcohol preference in rats lacking gustatory neocortex. *Alcohol, 4,* 37–43.

King, G. L. (1990). Animal models in the study of vomiting. *Canadian Journal of Physiology and Pharmacology, 68,* 260–268.

King, J. M. and Cox, V. C. (1976). Effects of ergocornine and CI-628 on food intake and taste preferences. *Physiology and Behavior, 16,* 111–113.

Kinney, F.C. (1978). An experimental study of the central gustatory pathways in the monkey. Macaca mulatta and Cercopithecus aethiops. *Journal fuer Hirnforschung, 19,* 21–43.

Kiyomitsu, Y. Yamamoto, T., Matsuo, R. and Kitamura, R. (1988). Centrifugal influence from the cortical gustatory area on neuronal activities of the

parabrachial nucleus in rats. *Journal of Physiological Society of Japan, 50,* 515.
Klein, S. B., Barter, M. J., Murphy, A. L., and Richardson, J. H. (1974). Aversion to low doses of mercuric chloride in rats. *Physiological Psychology, 2,* 397–400.
Kosar, E., Grill, H. J., and Norgren, R. (1986). Gustatory cortex in the rat. II. Thalamocortical connections. *Brain Research, 379,* 342–352.
Kossut, M. and Rose, S. P. R. (1984). Differential 2-deoxyglucose uptake into chick brain structures during passive avoidance training. *Neuroscience, 12,* 971–977.
Kosten, T., and Contreras, R. J. (1989). Deficits in conditioned heart rate and taste aversion in area postrema-lesioned rats. *Behavioural Brain Research, 35,* 9–21.
Kraemer, P. J., Randall, C. K., and Carbary, T. J. (1991). Release from latent inhibition with delayed testing. *Animal Learning and Behavior, 19,* 139–145.
Kramer, T. H., Sclafani, A., Kindya, K., and Pezner, M. (1983). Conditioned taste aversion in lean and obese rats with ventromedial hypothalamic knife cuts. *Behavioral Neuroscience, 97,* 110–119.
Krane, R. V., Sinnamon, H. M., and Thomas, G. J. (1976). Conditioned taste aversions and neophobia in rats with hippocampal lesions. *Journal of Comparative and Physiological Psychology, 90,* 680–693.
Krettek, J. E. and Price, J. L. (1978). Amygdaloid projections to subcortical structures within the basal forebrain and brainstem in the rat and cat. *Journal of Comparative Neurology, 178,* 225–254.
Krivanek, J. (1993). Protein kinase C in nucleus parabrachialis: effect of drugs inducing conditioned taste aversion. *Brain Research, 629,* 327–330.
Krivanek, J. (1996). Conditioned taste aversion and protein kinase C in the parabrachial nucleus of rats. *Neurobiology of Learning and Memory, 65,* 154–162.
Kromer, L. F., Björklund, A., and Stenevi, U. (1980). Innervation of embryonic hippocampal implants by regenerating axons of cholinergic septal neurons in the adult rat. *Brain Research, 210,* 153–171.
Krushel, L. A. and van der Kooy, D. (1988). Visceral cortex: integration of the mucosal senses with limbic information in the rat agranular insular cortex. *Journal of Comparative Neurology, 270,* 39–54.
Kumar, R., Pratt, J. A., and Stolerman, I. P. (1983). Characteristics of conditioned taste aversion produced by nicotine in rats. *British Journal of Pharmacology, 79,* 245–253.
Kupfermann, I. (1974). Feeding behavior in Aplysia: a simple system for the study of motivation. *Behavioral Biology, 10,* 101–126.
Kurihara, K., Suzuki, N., and Ogawa, H. (Eds.) (1994). *Olfaction and Taste, XI* (pp. 1–864). Tokyo: Springer.
Kuslansky, B., Weiss, K. R., and Kupfermann, I. (1987). Mechanisms underlying satiation of feeding behavior of the mollusc Aplysia. *Behavioral and Neural Biology, 48,* 278–303.
Kutscher, C. L. and Wright, W. A. (1977). Unconditioned taste aversion to quinine induced by injections of NaCl and LiCl: dissociation of aversion from cellular dehydration. *Physiology and Behavior, 18,* 87–94.

Labbe, R., Firl, A., Mutson, E. J., and Stein, O. G. (1983). Fetal brain transplants: reduction of cognitive deficit in rats with frontal cortex lesions. *Science, 221,* 470–472.

Ladowsky, R. L. and Ossenkopp, K. P. (1986). Conditioned taste aversions and changes in motor activity in lithium-treated rats. Mediating role of the area postrema. *Neuropharmacology, 25,* 71–77.

Lamon, S., Wilson, G. T., and Leaf, R. C. (1977). Human classical aversion conditioning: nausea versus electric shock in the reduction of target beverage consumption. *Behaviour Research and Therapy, 15,* 313–320.

Lamprecht, R. and Dudai, Y. (1995). Differential modulation of brain immediate early genes by intraperitoneal LiCl. *NeuroReport, 7,* 289–293.

Lamprecht, R. and Dudai, Y. (1996). Transient expression of c-*fos* in rat amygdala during training is required for encoding conditioned taste aversion memory. *Learning and Memory, 31,* 31–41.

Landauer, M. R., Lynch, M. R., Balster, R. L., and Kallman, M. J. (1982). Trichloromethane-induced taste aversions in mice. *Neurobehavioral Toxicology and Teratology, 4,* 305–309.

Lanteri-Minet, M., Isnardon, P., De Pommery, J., and Menetrey, D. (1993). Spinal and hindbrain structures involved in visceroception and visceronociception as revealed by the expression of Fos, Jun and Krox-24 proteins. *Neuroscience, 55,* 737–753.

Lapchak, P. A., Jenden, D. J., and Hefti, F. (1991). Compensatory elevation of acetylcholine synthesis *in vivo* by cholinergic neurons surviving partial lesions of the septohippocampal pathway. *Neuroscience, 11,* 2821–2828.

Lasiter, P. S. (1982). Cortical substrates of taste aversion learning: direct amygdalocortical projections to the gustatory neocortex do not mediate conditioned taste aversion learning. *Physiological Psychology 10,* 377–383.

Lasiter, P. S. (1985). Thalamocortical relations in taste aversion learning. II. Involvement of the medial ventrobasal thalamic complex in taste aversion learning. *Behavioral Neuroscience, 99,* 477–495.

Lasiter, P. S. and Glanzman, D. L. (1985). Cortical substrates of taste aversion learning: involvement of dorsolateral amygdaloid nuclei and temporal neocortex in taste aversion learning. *Behavioral Neuroscience, 99,* 257–276.

Lazlo, J. (1983). Nausea and vomiting as major complications of cancer chemotherapy, *Drugs, 25, Suppl. 1,* 1–6.

Leao, A. A. P. (1944). Spreading depression of activity in the cerebral cortex. *Journal of Neurophysiology, 7,* 359–390.

LeDoux, J. E. (1993). Emotional memory systems in the brain. *Behavioural Brain Research, 58,* 69–79.

LeDoux, J. E., Iwata, J., Cicchetti, P., and Reis, D. J. (1988). Different projections of the central amygdaloid nuclei mediate autonomic and behavioral correlates of conditioned fear. *Journal of Neuroscience, 8,* 2517–2529.

Lee, D. W. and Green, K. F. (1989). Effects of antihistamines on centrifugal rotation-induced analgesia and conditioned flavor aversions. *Physiology and Behavior, 45,* 459–464.

Lehr, P. P. and Nachman, M. (1973). Lateralization of learned taste aversion by

cortical spreading depression. *Physiology and Behavior, 10,* 79–83.
Le Magnen, J. (1985). *Hunger.* Cambridge: Cambridge University Press.
Levy, C. K., Ervin, F. R., and García, J. (1970). Effect of serum from irradiated rats on gastrointestinal function. *Nature, 225,* 463–464.
Levy, C. J., Carroll, M. E., Smith, J. C., and Hofer, K. G. (1974). Antihistamines block radiation-induced taste aversions. *Science, 186,* 1044–1046.
Lindsay, G. P. and Best, P. J. (1973). Overshadowing of the least salient of two novel fluids in a taste aversion paradigm. *Physiological Psychology, 1,* 13–15.
Logue, A. W. (1985). Conditioned food aversion learning in humans. *Annals of the New York Academy of Sciences, 443,* 316–329.
López-García, J. C., Bermúdez-Rattoni, F., and Tapia, R. (1990). Release of acetylcholine, gamma-aminobutyrate, dopamine and glutamate, and activity of some related enzymes, in rat gustatory neocortex. *Brain Research, 23,* 100–104.
López-García, J. C., Fernández-Ruíz, J., Bermúdez-Rattoni, F., and Tapia, R. (1990). Correlation between acetylcholine release and recovery of conditioned taste aversion induced by fetal neocortex grafts. *Brain Research, 523,* 105–110.
Lorden, J. F. and Nunn, W. B. (1982). Effects of central and peripheral pretreatment with fluoxetine in gustatory conditioning. *Pharmacology, Biochemistry and Behavior, 17,* 435–444.
Loullis, C. C., Wayner, M. J., and Jolicoeur, F. B. (1978). Thalamic taste nuclei lesions and taste aversion. *Physiology and Behavior, 20,* 653–655.
Macintosh, N. J. (1974). *The psychology of animal learning.* New York: Academic Press.
Mackay, B. (1977). Visual and flavor cues in toxicosis conditioning in the codfish. *Behavioral Biology, 19,* 87–97.
MacPhail, R. C. (1982). Studies on the flavor aversions induced by trialkyltin compounds. *Neurobehavioral Toxicology and Teratology, 4,* 225–230.
MacPhail, R. C. and Leander, J. D. (1980). Flavor aversions induced by chlordimeform. *Neurobehavioral Toxicology, 2,* 363–365.
Maier, S. F., Watkins, L. R., and Fleshner, M. (1994). Psychoneuroimmunology. The interface between behavior, brain and immunity. *American Psychology, 49,* 1004–1017.
Mallet, P. E., Beninger, R. J., Flesher, S. N., Jhamandas, K., and Boegman, R. J. (1995). Nucleus basalis lesions: implication of basoamygdaloid cholinergic pathways in memory. *Brain Research Bulletin, 36,* 51–56.
Mamedov, Z., Hernandez-Mesa, N., and Bures, J. (1987). Inefficient licking during forced spout alternation in rats: violation of the law of effect? *Physiology and Behavior, 39,* 153–160.
Maren, S., Poremba, A., and Gabrial, M. (1991). Basolateral amygdaloid multi-unit neuronal correlates of discriminative avoidance learning in rabbits. *Brain Research, 549,* 311–316.
Margules, D. L. (1970). Beta-adrenergic receptors in the hypothalamus for learned and unlearned taste aversions. *Journal of Comparative and Physiological Psychology, 73,* 13–21.

Martin, J. R., Cheng, F. Y., and Novin, D. (1978). Acquisition of learned taste aversion following bilateral subdiaphragmatic vagotomy in rats. *Physiology and Behavior, 21,* 13–17.

Maruniak, J. A., Mason, J. R., and Kostelc, J. G. (1983). Conditioned aversions to an intravascular odorant. *Physiology and Behavior, 30,* 617–620.

Mason, J. R. and Reidinger, R. F. (1983). Importance of color for methiocarb-induced food aversions in red-winged black-birds. *Journal of Wildlife Management, 47,* 383–393.

Mason, J. R., Rabin, M. D., and Stevens, S. D. (1982). Conditioned taste aversions: skin secretions used for defense by tiger salamanders, *Ambystoma tigrinum. Copeia, 3,* 667–671.

Matsuo, R. and Yamamoto, T. (1990). Taste nerve responses during licking behavior in rats: importance of saliva in responses to sweeteners. *Neuroscience Letters, 108,* 121–126.

Matsuo, R., Yamamoto, T., Ikehara, A., and Nakamura, O. (1994). Effects of salivation on neural taste responses in freely moving rats: analyses of salivary secretion and taste responses of the chorda tympani nerve. *Brain Research, 649,* 136–146.

McGowan, B., Hankins, W. G., and García, J. (1972). Limbic lesions and control of the internal and external environment. *Behavioral Biology, 7,* 841–852.

Meachum, C. L. and Bernstein, I. L. (1990). Conditioned responses to a taste conditioned stimulus paired with lithium chloride administration. *Behavioral Neuroscience, 104,* 711–715.

Menetrey, D., Gannon, A., Levine, J. D., and Basbaum A. I. (1989). Expression of c-fos protein in interneurons and projection neurons of the rat spinal cord in response to noxious somatic, articular, and visceral stimulation. *Journal of Comparative Neurology, 285,* 177–195.

Merril, J. E. and Jonakait, G. M. (1995). Interactions of the nervous and immune system in development, normal brain homeostasis, and disease. *FASEB Journal, 9,* 611–618.

Mesulam, M. M. (1982). *Tracing neural connections with horseradish peroxidase.* New York: John Wiley.

Miceli, D., Marfaing-Jallat, P., and Le Magnen, J. (1980). Ethanol aversion induced by parenterally administered ethanol acting both as a CS and UCS. *Physiological Psychology, 8,* 433–436.

Miller, D. B. and Miller, L. L. (1983). Bupropion, D-amphetamine, and amitriptyline-induced conditioned taste aversion in rats: dose effects. *Pharmacology, Biochemistry and Behavior, 18,* 737–740.

Miller, C. R., Elkins, R. L., and Peacock, L. J. (1971). Disruption of radiation-induced preference shift by hippocampal lesions. *Physiology and Behavior, 6,* 283–285.

Miller, C., Elkins, R., Fraser, J., Peacock, L., and Hobbs, S. (1975). Taste aversion and passive avoidance in rats with hippocampal lesions. *Physiological Psychology,* 3, 123–126.

Miller, J. S., Nonneman, A. J., Kelly, K. S., Neisewander, J. L., and Isaac, W. L. (1986). Disruption of neophobia, conditioned odor aversion, and conditioned taste aversion in rats with hippocampal lesions. *Behavioral and Neural Biology, 45,* 240–253.

Milner, B. (1972). Disorders of learning and memory after temporal lobe lesions in man. *Clinical Neurosurgery, 19,* 421–446.

Minano, F. J., Meneres Sancho, M. S., Sancibrian, M., Salinas, P., and Myers, R. D. (1992). GABA receptors in the amygdala: role in feeding in fasted and satiated rats. *Brain Research, 586,* 104–110.

Mineka, S., Seligman, M. E. P., Hetrick, M., and Zuelzer, K. (1972). Poisoning and conditioned drinking. *Journal of Comparative and Physiological Psychology, 79,* 377–383.

Mitchell, D., Wells, C., Hoch, N., Lind, K., Woods, S. C., and Mitchell, L. K. (1976). Poison induced pica in rats. *Physiology and Behavior, 17,* 681–697.

Miyagawa, M. (1982). Conditioned taste aversion induced by inhalation exposure to methyl bromide in rats. *Toxicology Letters, 10,* 411–416.

Moga, M. M. and Gray, T. S. (1985). Evidence for corticotropin-releasing factor, neurotensin, and somatostatin in neural pathway from the central nucleus of the amygdala to the parabrachial nucleus. *Journal of Comparative Neurology, 241,* 275–284.

Morgan, J. I. and Curran, T. (1986). Role of ion flux in the control of c-*fos* expression. *Nature, 322,* 552–555.

Morrison, G. R. and Jessup, A. (1977). A dual taste for saccharin in the rat. II. Taste change following alloxan injection. *Chemical Senses and Flavor, 2,* 395–400.

Mueller, K. and Hsiao, S. (1977). Specificity of cholecystokinin satiety in reduction of food but not water intake. *Pharmacology, Biochemistry and Behavior, 6,* 643–646.

Mueller, K. and Hsiao, S. (1978). Current status of cholecystokinin as a short term satiety hormone. *Neuroscience and Biobehavioral Reviews,* 2, 79–87.

Muramoto, K., Ono, T., Nishijo, H., and Fukuda, M. (1993). Rat amygdaloid neuron responses during auditory discrimination. *Neuroscience, 52,* 621–636.

Murphy, L. R. and Brown, T. S. (1974). Hippocampal lesions and learned taste aversion. *Physiological Psychology, 2,* 60–64.

Myers, R. H. and DeCastro, J. M. (1977). Learned aversion to intracerebral carbachol. *Physiology and Behavior, 19,* 467–472.

Myhal, N. and Fleming, A. (1990). MK-801 effects on a learned food preference depends on dosage: is it disruption of learning or a conditioned aversion? *Psychobiology, 18,* 428–434.

Nachman, M. (1963). Learned aversion to the taste of lithium chloride and generalization to other salts. *Journal of Comparative and Physiological Psychology, 56,* 343–349.

Nachman, M. (1970). Learned taste and temperature aversions due to lithium chloride sickness after temporal delays. *Journal of Comparative and Physiological Psychology, 73,* 22–30.

Nachman, M. and Ashe, J .H. (1973). Learned taste aversions in rats as a function of dosage, concentration and route of administration of LiCl. *Physiology and Behavior, 10,* 73–78.

Nachman, M. and Ashe, J. H. (1974). Effects of basolateral amygdala lesions on neophobia, learned taste aversions, and sodium appetite in rats. *Journal of Comparative and Physiological Psychology, 87,* 622–643.

Nachman, M. and Hartley, P. L. (1975). Role of illness in producing learned taste aversions in rats: a comparison of several rodenticides. *Journal of Comparative and Physiological Psychology, 89,* 1010–1018.

Nachman, M. and Jones, D. R. (1974). Learned taste aversions over long delays in rats: the role of learned safety. *Journal of Comparative and Physiological Psychology, 86,* 949–956.

Nachman, M., Lester, D., and Le Magnen, J. (1970). Alcohol aversion in the rat: behavioral assessment of noxious drug effects. *Science, 168,* 1244–1246.

Nakajima, S. (1974). Conditioned aversion of water produced by cycloheximide injection. *Physiological Psychology, 2,* 484–486.

Naor, C. and Dudai, Y. (1996). Transient impairment of cholinergic function in the rat insular cortex disrupts the encoding of taste in conditioned taste aversion. *Behavioural Brain Research, 79,* 61–67.

Nauta, W. J. H. (1971). The problem of the frontal lobe: a reinterpretation. *Journal of Psychiatric Research, 8,* 167–187.

Neafsey, E. J., Sorensen, J. C., Tonder, N., and Castro A. J. (1988). Fetal cortical transplants into neonatal rats respond to thalamic and peripheral stimulation in the adult. An electrophysiological study of single-unit activity. *Brain Research, 493,* 33–40.

Nerad, L., Ramírez-Amaya, V., Ormsby, C., and Bermúdez-Rattoni, F. (1996). Differential effects of anterior and posterior insular cortex lesions on the acquisition of conditioned taste aversion and spatial learning. *Neurobiology of Learning and Memory, 65,* 1–7.

Nieto-Sampedro, M. and Cotman, C. W. (1986). Growth factor induction and other events in CNS repair. In C.W. Cotman (Ed.) *Synaptic plasticity and remodeling* (pp. 407–456). New York: Gilford Press.

Niijima, A. and Yamamoto, T. (1994). The effects of lithium chloride on the activity of the afferent nerve fibers from the abdominal visceral organs in the rat. *Brain Research Bulletin, 35,* 141–145.

Ninomiya, Y., Higashi, T., Katsukawa, H., Mizukoshi, T., and Funakoshi, M. (1984). Qualitative discrimination of gustatory stimuli in three different strains of mice. *Brain Research, 322,* 83–92.

Nonneman, A. J. and Curtis, S. D. (1978). Strength of conditioning determines the effects of septo-hippocampal lesions on taste aversion learning. *Physiological Psychology, 6,* 249–254.

Norgren, R. (1974). Gustatory afferents to ventral forebrain. *Brain Research, 81,* 285–295.

Norgren, R. (1976). Taste pathways to hypothalamus and amygdala. *Journal of Comparative Neurology, 166,* 17–30.

Norgren, R. (1978). Projections from the nucleus of the solitary tract in the rat. *Neuroscience, 3,* 207–218.

Norgren, R. (1984). Central neural mechanisms of taste. In I. Darian-Smith (Ed.), *Handbook of physiology, Section I, The nervous system, Vol. III, Sensory processes, Part 2* (pp. 1087–1128). Bethesda: American Physiological Society.

Norgren, R. and Grill, H. J. (1976). Efferent distribution from the cortical gustatory area in rats. *Neuroscience Society Abstracts, 2,* 124.

Norgren, R. and Leonard, C. M. (1971). Taste pathways in rat brainstem. *Science, 173,* 1136–1139.

Norgren, R. and Wolf, G. (1975). Projections of thalamic gustatory and lingual areas in the rat. *Brain Research, 92,* 123–129.

Nowlis, G. H., Frank, M. E., and Pfaffmann, C. (1980). Specificity of acquired aversions to taste qualities in hamsters and rats. *Journal of Comparative and Physiological Psychology, 94,* 932–942.

Ogawa, H., Hayama, T., and Ito, S. (1982). Convergence of input from tongue and palate to the parabrachial nucleus neurons of rats. *Neuroscience Letters, 28,* 15–20.

Ogawa, H., Ito, S., Murayama, N., and Hasegawa, K. (1990). Taste area in granular and dysgranular insular cortices in the rat identified by stimulation of the entire oral cavity. *Neuroscience Research, 9,* 196–210.

Ono, T. and Nishijo, H. (1992). Neurophysiological basis of the Kluver Bucy syndrome: responses of monkey amygdaloid neurons to biologically significant objects. In J. P. Aggleton (Ed.), *The amygdala: neurobiological aspects of emotion, memory, and mental dysfunction* (pp. 167–190). New York: Wiley Liss.

Oomura, Y. and Ono, T. (1982). Mechanism of inhibition by the amygdala in the lateral hypothalamic area of the rats. *Brain Research Bulletin, 8,* 653–666.

Ormsby, C. E., Piña A. L., and Bermúdez-Rattoni, F. (1991). Long-term retrograde amnesia of inhibitory avoidance and conditioned taste aversion learning tasks by insular cortex lesions. *Society for Neuroscience Abstracts, 17,* 1045.

Ormsby, C. E., Ramírez-Amaya, V., and Bermúdez-Rattoni, F. (1994). Insular cortex grafts restore rememberance of previously learned taste aversions. *Society for Neuroscience Abstracts, 20,* 1211.

Ormsby, C. E., González-Rodríguez, C. L., and Bermúdez-Rattoni, F. (1996). Dorsal insular cortex lesions impair the recall, but not the acquisition of the Morris water maze. *Society for Neuroscience Abstracts, 22,* 682.

Ossenkopp, K.-P. (1983*a*). Taste aversions conditioned with gamma radiation: attenuation by area postrema lesions in rats. *Behavioural Brain Research, 7,* 297–305.

Ossenkopp, K.-P. (1983*b*). Area postrema lesions in rats enhance the magnitude of body rotation-induced conditioned taste aversions. *Behavioral and Neural Biology, 38,* 82–96.

Ossenkopp, K.-P. and Eckel, L. A. (1995). Toxin induced conditioned changes in taste reactivity and the role of the chemosensitive area postrema. *Neuroscience and Biobehavioral Reviews, 19,* 99–108.

Ossenkopp, K.-P. and Giugno, L. (1985). Taste aversions conditioned with multiple exposures to gamma radiation: abolition by area postrema lesion in rats. *Brain Research, 346,* 1–7.

Ossenkopp, K.-P. and Giugno, L. (1990). Nicotine-induced conditioned taste aversions are enhanced in rats with lesions of the area postrema. *Pharma-*

cology Biochemistry and Behavior, 36, 625–630.

Ossenkopp, K.-P., Sutherland, C., and Ladowsky, R. L. (1986). Motor activity changes and conditioned taste aversions induced by administration of scopolamine in rats: role of the area postrema. *Pharmacology Biochemistry and Behavior, 25,* 269–276.

Ottersen, O. P. and Ben-Ari, Y. (1979). Afferent connections to the amygdaloid complex of the rat and cat. I. Projections from the thalamus. *Journal of Comparative Neurology, 187,* 401–424.

Palmerino, C. C., Rusiniak, K. W., and García, J. (1980). Flavor–illness aversions: the peculiar roles of odor and taste in memory for poison. *Science, 208,* 753–755.

Pan, P. Q. and Long, J. (1993). Lesions of the hippocampus enhance or depress humoral immunity in rats. *NeuroReport, 4,* 864–866.

Panksepp, J., Pollack, A., Meeker, R. B., and Sullivan, A. C. (1977). (-)-Hydroxycitrate and conditioned aversions. *Pharmacology, Biochemistry and Behavior, 6,* 683–687.

Parker, L. A. (1980). Conditioned suppression of drinking: a measure of the CR elicited by a lithium-conditioned flavor. *Learning and Motivation, 11,* 538–559.

Parker, L. A. (1982). Nonconsummatory and consummatory behavioral CRs elicited by lithium- and amphetamine-paired flavors. *Neurobehavioral Toxicology and Teratology, 5,* 23–27.

Parker, L. A. (1995). Rewarding drugs produce taste avoidance, but not taste aversion. *Neuroscience and Biobehavioral Reviews, 19,* 143–151.

Parker, L. A. and Revusky, S. (1982). Generalized conditioned flavor aversions: effects of toxicosis training with one flavor on the preference for different novel flavors. *Animal Learning and Behavior, 10,* 505–510.

Pascoe, J. P. and Kapp, B. S. (1987). Responses of amygdaloid central nucleus neurons to stimulation of the insular cortex in awake rabbits, *Neuroscience, 21,* 471–485.

Patterson, T. A., Rose, S. P. R., and Bradley, P. M. (1989). Anisomycin and amnesia in the chick: state-dependent effects are not present with intracranial injections. *Developmental Brain Research, 49,* 173–178.

Pelchat, M. L., Grill, H. J., Rozin, P., and Jacobs, J. (1983). Quality of acquired responses to tastes by Rattus norvegicus depends on types of associated discomfort. *Journal of Comparative and Physiological Psychology, 97,* 140–153.

Pfaffmann, C. (1941). Gustatory afferent impulses. *Journal of Cellular and Comparative Physiology, 17,* 243–258.

Pfaffmann, C. (1974). Specificity of the sweet receptors of the squirrel monkey. *Chemical Senses, 1,* 61–67.

Phillips, A. G. and Le Piane, F. G. (1978). Electrical stimulation of the amygdala as a conditioned stimulus in a bait-shyness paradigm. *Science, 201,* 536–538.

Phillips, A. G. and Le Piane, F. G. (1980). Disruption of conditioned taste aversion in the rat by stimulation of amygdala: a conditioning effect, not amnesia. *Journal of Comparative and Physiological Psychology, 94,* 664–674.

Poulton, E. B. (1887). The experimental proof of the protective value of color and marking in insects in reference to their vertebrate enemies. *Proceedings of the Zoological Society of London*, 191–274.

Preston, K. L., Wagner, G. C., Seiden, L. S., and Schuster, C. R. (1984). Effects of methamphetamine on atropine-induced conditioned gustatory avoidance. *Pharmacology, Biochemistry and Behavior, 20,* 601–607.

Price, J. L. and Amaud, D. G. (1981). An autoradiographic study of the projections of the central nucleus of the monkey amygdala. *Journal of Neuroscience, 1,* 1242–1259.

Rabin, B. M. and Rabin, J. S. (1984). Acquisition of radiation- and lithium chloride-induced conditioned taste aversions in anesthetized rats. *Animal Learning and Behavior, 12,* 439–441.

Rabin, B. M., Hunt, W. A., and Lee, J. (1983*a*). Attenuation of radiation- and drug-induced conditioned taste aversions following area postrema lesions in the rat. *Radiation Research, 93,* 388–394.

Rabin, B. M., Hunt, W. A., and Lee, J. (1983*b*). Recall of a previously acquired conditioned taste aversion in rats following lesions of the area postrema. *Physiology and Behavior, 32,* 119–122.

Rabin, B. M., Hunt, W. A., and Lee, J. (1987). Interactions between radiation and amphetamine in taste aversion learning and the role of the area postrema in amphetamine-induced conditioned taste aversions. *Pharmacology Biochemistry and Behavior, 27,* 677–683.

Ramírez-Amaya, V., Alvarez-Borda, B., Ormsby, C. E., Martínez, R. D., Pérez-Montfort, R., and Bermúdez-Rattoni, F. (1996). Insular cortex lesions impair the acquisition of conditioned immunosuppression. *Brain, Behavior and Immunity, 10,* 103–114.

Rappold, V. A., Porter, J. H., and Llewellyn, G. C. (1984). Evaluation of the toxic effects of aflatoxin B1 with a taste aversion paradigm in rats. *Neurobehavioral Toxicology and Teratology, 6,* 51–58.

Reilly, S. and Pritchard, T. C. (1996). Gustatory thalamus lesions in the rat: U. Aversive and appetitive taste conditioning. *Behavioral Neuroscience, 110,* 746–759.

Reilly, S., Grigson, P. S., and Norgren, R. (1993*a*). Parabrachial nucleus lesions and conditioned taste aversion: evidence supporting an associative deficit. *Behavioral Neuroscience, 107,* 1005–1017.

Reilly, S., Harley, C., and Revusky, S. (1993*b*). Ibotenate lesions of the hippocampus enhance latent inhibition in conditioned taste aversion and increase resistance to extinction in conditioned taste preference. *Behavioral Neuroscience, 107,* 996–1004.

Revusky, S. (1968). Aversion to sucrose produced by contingent X-irradiation: temporal and dosage parameters. *Journal of Comparative and Physiological Psychology, 65,* 17–22.

Revusky, S. H. (1971). The role of interference in association over a delay. In W. Honig and H. James (Eds.), *Animal memory* (pp. 155–213). New York: Academic Press.

Revusky, S. (1977). Interference with progress by the scientific establishment:

examples from taste aversion learning. In N. Milgram, L. Krames, and T. Alloway (Eds.), *Food aversion learning* (pp. 53–71). New York: Plenum Press.

Revusky, S. and Gorry, T. (1973). Flavor aversions produced by contingent drug injection: relative effectiveness of apomorphine, emetine, and lithium. *Behaviour Research and Therapy, 11,* 403–409.

Ricardo, J. A. and Koh, E. T. (1978). Anatomical evidence of direct projections from the nucleus of the solitary tract to the hypothalamus, amygdala, and other forebrain structures in the rat. *Brain Research, 153,* 1–26.

Rice, A. G. (1988). Estrogen produces conditioned taste aversions: Doctoral Thesis, Psychology Department, University of California, Los Angeles, 90024.

Rice, A. G., López, A., and García, J. (1987). Estrogen produces conditioned taste aversions in rats which are blocked by antihistamine. *Society for Neuroscience Abstracts, 13,* 556.

Richardson, S. J. (1973). The amygdala: historical and functional analysis. *Acta Neurobiologiae Experimentalis, 33,* 623–648.

Richter, C. P. (1943). Total selfregulatory function in animals and human beings. *Harvey Lectures Series, 38,* 63–103.

Riley, A. L. and Tuck, D. L. (1985). Conditioned taste aversions: a bibliography. *Annals of the New York Academy of Sciences, 443,* 381–437.

Riley, A. L. and Zellner, D. A. (1978). Methylphenidate-induced conditioned taste aversion: an index of toxicity. *Physiological Psychology, 6,* 354–358.

Rigter, H. and Popping, A. (1975). Hormonal influences on the extinction of conditioned taste aversion. *Psychopharmacologia, 46,* 255–261.

Ritter, S., McGlone, J. J., and Kelley, K. W. (1980). Absence of lithium-induced taste aversion after area postrema lesion. *Brain Research, 201,* 501–506.

Robertson, R. T., Hanes, M. A., and Yu, J. (1988). Investigations of the origins of transient acetyl-cholinesterase activity in developing rat visual cortex. *Developmental Brain Research, 41,* 1–23.

Robertson, R. T., Mostamand, F., Kageyama, G. H., Gallardo, K. A., and Yu, J. (1991). Primary auditory cortex in the rat: transient expression of acetylcholinesterase activity in developing geniculocortical projections. *Developmental Brain Research, 58,* 81–95.

Rodgers, W. L. (1967). Specificity of specific hungers. *Journal of Comparative and Physiological Psychology, 64,* 49–58.

Roldan, G. and Bures, J. (1994). Tetrodotoxin blockade of amygdala overlapping with poisoning impairs acquisition of conditioned taste aversion in rats. *Behavioural Brain Research, 65,* 213–219.

Roll, D. L. and Smith, J. C. (1972). Conditioned taste aversion in anesthetized rats. In M. E. P. Seligman and J. L. Hager (Eds.), *Biological boundaries of learning* (pp. 98–102). New York: Appleton-Century-Crofts.

Rolls, E. T. (1975). *The brain and reward.* Oxford: Pergamon Press.

Rolls, B. J. and Rolls, E. T. (1973). Effects of lesions in the basolateral amygdala on fluid intake in the rat. *Journal of Comparative and Physiological Psychology, 83,* 240–247.

Room, P. and Groenewegen, H. J. (1986). Connections of the parahippocampal cortex. I. Cortical afferents. *Journal of Comparative Neurology, 251,* 415–450.

Roozendaal, B., Oldenburger, W. P., Strubble, J. H., Koolhaas, J. M., and Bohus, B. (1990). The central amygdala is involved in the conditioned but not in the meal-induced cephalic insulin response in the rat. *Neuroscience Letters, 116,* 210–215.

Rose, S. P. R. (1995). Memory for taste aversion: molecular and cellular mechanisms. In K. Kurihara, N. Suzuki, and H. Ogawa (Eds.), *Olfaction and taste, XI* (pp. 462–466). Tokyo: Springer Verlag.

Rosenblum, K., Meiri, N., and Dudai, Y. (1993). Taste memory: the role of protein synthesis in gustatory cortex. *Behavioral and Neural Biology, 59,* 49–56.

Rosenblum, K., Schul, R., Meiri, N., Hadari, Y.R., Zick, Y., and Dudai, Y. (1995). Modulation of protein tyrosine phosphorylation in rat insular cortex after conditioned taste aversion training. *Proceedings of the National Academy of Sciences USA, 92,* 1157–1161.

Ross, D. T. and Ebner, F. F. (1985). Neocortical transplants in the S1 cortex of adult mice receive thalamic afferents from the posteromedial (POm) but not the ventrobasal (VB) nucleus. *Society for Neuroscience Abstracts, 11,* 974.

Roszman, T. L., Cross, R. L., Brooks, W. H., and Markesberry, W. R. (1982). Hypothalamic immune interactions II. *Immunology, 45,* 737–742.

Roth, S., Schwartz, M., and Teitelbaum, P. (1973). Failure of recovered lateral hypothalamic rats to learn specific food aversions. *Journal of Comparative and Physiological Psychology, 83,* 184–197.

Rozin, P. (1968). Specific aversions and neophobia resulting from vitamin deficiency or poisoning in half-wild and domestic rats. *Journal of Compartive and Physiological Psychology, 66,* 82–88.

Rozin, P. and Kalat, J. (1971). Specific hungers and poison avoidance as adaptive specializations of learning. *Psychological Reviews, 78,* 459–486.

Rozin, P. and Ree, P. (1972). Long extension of effective CS–US interval by anesthesia between CS and US. *Journal of Comparative and Physiological Psychology, 80,* 43–48.

Rusiniak, K. W., Hankins, W. G., García, J., and Brett, L. P. (1979). Flavor–illness aversions: potentiation of odor by taste in rats. *Behavioral and Neural Biology, 25,* 1–17.

Rusiniak K. W., Palmerino, C. C., Rice, A. G., Forthman, D. L., and García, J. (1982). Flavor–illness aversions: potentiation of odor by taste with toxin but not shock in rats. *Journal of Comparative and Physiological Psychology, 96,* 527–539.

Russell, R., Escobar, M. L. Booth, R. A., and Bermúdez-Rattoni, F. (1994). Nerve growth factor accelerates functional recovery in cortical lesioned rats. II. *In vivo* evidence for cholinergic involvement. *Behavioral and Neural Biology, 61,* 81–92.

Rzoska, J. (1953). Bait shyness, a study in rat behaviour. *British Journal of Animal Behaviour, 1,* 128–135.

Sahley, C., Gelperin, A., and Rudy, J. W. (1981). One-trial associative learning modifies food odor preference of terrestrial mollusc. *Proceedings of the National Academy of Sciences USA, 78,* 640–642.

Sakai, N., and Yamamoto, T. (1997). Conditioned taste aversion and c-*fos* expression in the rat brainstem after administration of various USs. *NeuroReport, 8,* 2215–2220.

Sakai, R. R., Fine, W. B., and Epstein A. N. (1987). Salt appetite is enhanced by one prior episode of sodium depletion in the rat. *Behavioral Neuroscience, 101,* 724–731.

Sakai, N., Tanimizu, T., Sako, N., Shimura, T., and Yamamoto, T. (1994). Effects of lesions of the medial and lateral parabrachial nuclei on acquisition and retention of conditioned taste aversion. In K. Kurihara, N. Suzuki, and H. Ogawa (Eds.), *Olfaction and taste, XI* (pp. 495–496). Tokyo: Springer-Verlag.

Saleh, T. M. and Cechetto, D. F. (1993). Peptides in the parabrachial nucleus modulate visceral input to the thalamus. *American Journal of Physiology, 264,* R668-R675.

Sandner, G., Di Scala, G., Oberling, P., Abbe, J. C., Stampfler, A., and Sens, J. C. (1994). Distribution of lithium in the rat brain after a single administration known to elicit aversive effects. *Neuroscience Letters, 166,* 1–4.

Saper, C. B. (1982*a*). Convergence of autonomic and limbic projections in the insular cortex of the rat. *Journal of Comparative Neurology, 2,* 163–173.

Saper, C. B. (1982*b*). Reciprocal parabrachial cortical connections in the rat. *Brain Research, 242,* 33–40.

Saper, C. B. and Loewy, A. D. (1980). Efferent connections of the parabrachial nucleus in the rat. *Brain Research, 197,* 291–317.

Schafe, G. E., Sollars, S. I. and Bernstein, I. L. (1995*a*). The CS–US interval and taste aversion learning: a brief look. *Behavioral Neuroscience, 109,* 799–802.

Schafe, G. E., Seeley, R. J., and Bernstein, I. L. (1995*b*). Forebrain contribution to the induction of a cellular correlate of conditioned taste aversion in the nucleus of the solitary tract. *Journal of Neuroscience, 15,* 6789–6796.

Schlessinger, J. and Ullrich, A. (1992). Growth factor signaling by receptor tyrosine kinases. *Neuron, 9,* 383–391.

Schulkin, J. (1991). *Sodium hunger: the search for a salty taste.* Cambridge: Cambridge University Press.

Schwarz, M. and Susswein, A. J. (1986). Identification of the neural pathway for the reinforcement of feeding when Aplysia learn that food is inedible. *Journal of Neuroscience, 6,* 1528–1536.

Schwartz, M. and Teitelbaum, P. (1974). Dissociation between learning and remembering in rats with lesions in the lateral hypothalamus. *Journal of Comparative and Physiological Psychology, 87,* 384–398.

Schwarz, M., Markovich, S., and Susswein, A. J. (1988). Parametric features of inhibition of feeding in Aplysia by associative learning, satiation and sustained lip stimulation. *Behavioral Neuroscience, 102,* 124–133.

Schwarz, M., Feldman, E., and Susswein, A. J. (1991). Variables affecting long-term memory of learning that food is inedible in Aplysia. *Behavioral Neuroscience, 105,* 194–201.

Scott, T. R. (1974). Behavioral support for a neural taste theory. *Physiology and Behavior, 12,* 413–417.

Scott, T. R. and Giza, B. K. (1987). The effect of physiological condition on taste in rats and primates. In Y. Kawamura and M. R. Kare (Eds.), *Umami: a basic taste* (pp. 409–437). New York: Marcel Dekker.

Seligman, M. E. P. (1970). On the generality of the laws of learning. *Psychological Review, 77,* 406–418.

Seligman, M. E. P. and Hager, J. L. (1972). *Biological boundaries of learning.* New York: Appleton-Century-Crofts.

Seward, J. P. and Greathouse, S. R. (1973). Appetitive and aversive conditioning in thiamine-deficient rats. *Journal of Comparative and Physiological Psychology, 83,* 157–167.

Shapiro, R. E. and Miselis, R. R. (1985). The central neural connections of the area postrema of the rat. *Journal of Comparative Neurology, 234,* 344–364.

Sharp, F. R. and González, M. F. (1986). Fetal cortical transplants ameliorate thalamic atrophy ipsilateral to neonatal frontal cortex lesions. *Neuroscience Letters, 71,* 247–251.

Shimura, T., Suzuki, M., and Yamamoto, T. (1995). Aversive taste stimuli facilitate extracellular acetylcholine release in the insular gustatory cortex of the rat: a microdialysis study. *Brain Research, 679,* 221–226.

Shimura, T., Grigson, P. S., and Norgren, R. (1997*a*). Brainstem lesions and gustatory function: I. The role of the nucleus of the solitary tract during a brief intake test in rats. *Behavioral Neuroscience, 111,* 155–168.

Shimura, T., Tanaka, H., and Yamamoto, T. (1997*b*). Salient responsiveness of parabrachial neurons to the conditioned stimulus after the acquisition of taste aversion learning in rats. *Neuroscience, 81,* 239–247.

Shipley, M. T. and Sanders, M. T. (1982). Special senses are really special: evidence for reciprocal bilateral pathway between insular cortex and nucleus parabrachialis. *Brain Research Bulletin, 8,* 493–501.

Sinden, J. D., Hodges, H., and Gray, J. A. (1995). Neural transplantation and recovery of cognitive function. *Behavioral and Brain Sciences, 18,* 10–35.

Sjoden, P.-O., Archer, T., and Carter, N. (1979). Conditioned taste aversion induced by 2,4,5-trichlorophenoxyacetic acid: dose-response and preexposure effects. *Physiological Psychology, 7,* 93–96.

Smith, G. P., Gibbs, J., and Kulkosky, P. J. (1982). Relationships between brain–gut peptides and neurons in the control of food intake. In B. Hoebel and D. Novin (Eds.), *The neural basis of feeding and reward* (pp. 149–256). Brunswick, ME: Haer Institute.

Smotherman, W. P. and Levine, S. (1978). ACTH and ACTH 4–10 modification of neophobia and taste aversion responses in the rat. *Journal of Comparative and Physiological Psychology, 92,* 22–33.

Smotherman, W. P., Hennessy, J. W., and Levine, S. (1976). Plasma corticosterone levels during recovery from LiCl produced taste aversions. *Behavioral Biology, 16,* 401–412.

Spector, A. C., Breslin, P., and Grill, H. J. (1988). Taste reactivity as a dependent measure of the rapid formation of conditioned taste aversion: a tool for the neural analysis of taste–visceral associations. *Behavioral Neuroscience, 102,* 942–952.

Spector, A. C., Schwartz, G. J., and Grill, H. J. (1990). Chemospecific deficits in taste detection after selective gustatory deafferentation in rats. *American Journal of Physiology, 258,* R820–R826.

Spector, A. C., Norgren, R., and Grill, H. J. (1992). Parabrachial gustatory lesions impair taste aversion learning in rats. *Behavioral Neuroscience, 106,* 147–161.

Sperry, R. W. (1961). Cerebral organization and behavior. *Science, 133,* 1749–1757.

Squire, L. R., Emanuel, C. A., Davis, H. P., and Deutsch, J. A. (1975). Inhibitors of cerebral protein synthesis: dissociation of aversive and amnesic effects. *Behavioral Biology, 14,* 335–341.

Staddon, J. E. R. (1983). *Adaptive behaviour and learning.* Cambridge: Cambridge University Press.

Stellar, E. (1954). The physiology of motivation. *Psychological Review, 61,* 5–22.

Stephens, D. W. and Krebs, J. R. (1986). *Foraging theory.* New Jersey: Princeton University Press.

Stewart, R. B., Perlanski, E., and Grupp, L. A. (1988). Area postrema and alcohol: effects of area postrema lesions on ethanol self-administration, pharmacokinetics, and ethanol-induced conditioned taste aversion. *Alcoholism: Clinical and Experimental Research, 12,* 698–704.

Stoerig, P. and Cowey, A. (1995). Visual perception and phenomenal consciousness. *Behavioural Brain Research, 71,* 147–156.

Stolerman, I. P. and Kumar, R. (1972). Secondary reinforcement in opioid dependence. In J. M. Sinngh, L. H. Miller, and H. Lai (Eds.), *Drug addiction: experimental psychology, Vol. 11* (pp. 49–60). New York: Futura.

Stolerman, I. P., Pilcher, C. W. T., and D'Mello, G. D. (1978). Stereospecific aversive property of narcotic antagonists in morphine-free rats. *Life Sciences, 22,* 1755–1762.

Sturdevant, R. A. L. and Goetz, H. (1976). Cholecystokinin both stimulates and inhibits human food intake. *Nature, 261,* 713–715.

Suemori, K., Kobashi, M., and Adachi, A. (1994). Effects of gastric distention and electrical stimulation of dorsomedial medulla on neurons in parabrachial nucleus of rats. *Journal of Autonomic Nervous System, 48,* 221–229.

Sugiura, Y., Terui, N., and Hosoya, Y. (1989). Difference in distribution of central terminals between visceral and somatic unmyelinated (C) primary afferent fibers. *Journal of Neurophysiology, 62,* 834–840.

Sutton, R. L., Fox, R. A., and Daunton, N. G. (1988). Role of the area postrema in three putative measures of motion sickness in the rat. *Behavioral and Neural Biology, 50,* 133–152.

Swank, M. W. and Bernstein, I. L. (1994). c-*fos* induction in response to a conditioned stimulus after single trial taste aversion learning. *Brain Research, 636,* 202–208.

Swank, M. W., Schafe, G. E., and Bernstein, I. L. (1995). c-*fos* induction in response to taste stimuli previously paired with amphetamine or LiCl during taste aversion learning. *Brain Research, 673,* 251–261.

Swank, M. W., Ellis, A. E., and Cochran, B. N. (1996). c-*fos* antisense blocks acquisition and extinction of conditioned taste aversion in mice. *NeuroReport, 7,* 1866–1870.

Swerdlow, N. R., van der Kooy, C., Koob, G. F., and Wenger, J. R. (1983). Cholecystokinin produces conditioned place-aversions, not place-preferences, in food-deprived rats: evidence against involvement in satiety. *Life Science, 32,* 2087–2093.

Thiele, T. E., Roitman, M. F., and Bernstein, I. L. (1996). c-*fos* induction in rat brainstem in response to ethanol- and lithium chloride-induced conditioned taste aversion. *Alcoholism: Clinical and Experimental Research, 20,* 1023–1028.

Thomka, M. L. and Brown, T. S. (1975). The effect of hippocampal lesions on the development and extinction of a learned taste aversion for a novel food. *Physiological Psychology, 3,* 281–284.

Thompson, R. (1983). Brain systems and long-term memory. *Behavioral and Neural Biology, 37,* 1–45.

Thompson, C. I. and Zagon, I. S. (1981). 2-Deoxy-D-glucose produces delayed hypophagia and conditioned taste aversion in rats. *Physiology and Behavior, 27,* 1001–1004.

Toates, F. (1986). *Motivational systems.* Cambridge: Cambridge University Press.

Todd, J. W. and Kesner, R. P. (1978). Effects of post-training injections of cholinergic agonists and antagonists into the amygdala on retention of passive avoidance in rats. *Journal of Comparative and Physiological Psychology, 92,* 958–968.

Torvik, A. (1956). Afferent connections to the sensory trigeminal nuclei, the nucleus of the solitary tract and adjacent structures. *Journal of Comparative Neurology, 106,* 51–132.

Toumane, A. and Durkin, T. P. (1993). Time gradient for post-test vulnerability to scopolamine-induced amnesia following the initial acquisition session of a spatial reference memory task in mice. *Behavioral and Neural Biology, 60,* 139–151.

Traub, R. J., Lim, F., Sengupta, J. N., Meller, S. T., and Gebhart, G. F. (1994). Noxious distention of viscera results in differential c-*fos* expression in second order sensory neurons receiving 'sympathetic' or 'parasympathetic' input. *Neuroscience Letters, 180,* 71–75.

Travers, S. P., Becker, D. C., Halsell, C. B., and Harper, M. I. (1994). Functional organization of the orally responsive NST. In K. Kurihara, N. Suzuki, and H. Ogawa (Eds.), *Olfaction and taste, XI* (pp. 396–401). Tokyo: Springer-Verlag.

Treneer, C. M. and Bernstein, I. L. (1983). Learned aversions in rats fed a tryptophan-free diet. *Physiology and Behavior, 27,* 757–760.

Tsukamoto, G. and Adachi, A. (1994). Neural responses of rat area postrema to stimuli producing nausea. *Journal of Autonomic Nervous System, 49,* 55–60.

Tucker, A. and Gibbs, M. (1976). Cycloheximide-induced amnesia for taste aversion memory in rats. *Pharmacology, Biochemistry and Behavior, 4,* 181–184.

Tucker, A. R. and Gibbs, M. E. (1979). Saccharin aversion memory in rats: inhibition of cycloheximide-resistant memory by ouabain. *Physiology and Behavior, 23,* 341–346.

Tucker, A. R. and Oei, T. S. (1982). Protein synthesis inhibition and amnesia for saccharin aversion memory in rats after intra-cisternal administration of cycloheximide. *Physiology and Behavior, 28,* 1025–1028.

Vahlsing, H. L., Hagg, T., Spencer, M., Conner, J. M., Manhorpe, M., and Varon, S. (1991). Dose-dependent responses to nerve growth factor by adult rat cholinergic medial septum and neostriatal neurons. *Brain Research, 522,* 320–329.

van der Kooy, D. and Koda, L. Y. (1983). Organization of the projections of a circumventricular organ: the area postrema in the rat. *Journal of Comparative Neurology, 219,* 328–338.

van der Kooy, D., Swerdlow, N. R., and Koob, G. F. (1983). Paradoxical reinforcing properties of apomorphine: effects of nucleus accumbens and area postrema lesions. *Brain Research, 259,* 2087–2094.

van der Kooy, D. L., Koda, L. Y., McGinty, C. R., Gerfen, C. R., and Bloom, F. E. (1984). The organization of projections from the cortex, amygdala and hypothalamus to the nucleus of the solitary tract in rat. *Journal of Comparative Neurology, 224,* 1–24.

Varon, S., Hagg, T., Vahlsing, H. L., and Manthorpe, M. (1989). Nerve growth factor in *in vivo* actions on cholinergic neurons in the adult rat CNS. In E. Cañedo, L. E. Todd, L. Packer, and J. Jaz (Eds.), *Cell function and disease* (pp. 235–248). New York: Plenum Press.

Vogel, J. R. and Nathan, B. A. (1975). Learned taste aversions induced by high doses of monosodium L-glutamate. *Pharmacology, Biochemistry and Behavior, 3,* 935–937.

Vogel, J. R. and Nathan, B. A. (1976). Reduction of learned taste aversions by preexposure to drugs. *Psychopharmacology, 49,* 167–172.

Waldbilling, R. J. and Callahan, M. O. (1980). Hormones and hedonics. Cholecystokinin and taste: a possible behavioral mechanism of action. *Physiology and Behavior, 25,* 25–30.

Weijnen, J. A. W. M. (1977). Current licking: lick contingent electrical stimulation of tongue. In J. Weijnen and J. Mendelson (Eds.), *Drinking behavior. Oral stimulation, reinforcement and preference.* New York: Plenum Press.

Weijnen, J. A. W. M. and Mendelson, J. (Eds.) (1977). *Drinking behavior. Oral stimulation, reinforcement and preference.* New York: Plenum Press.

Weiskrantz, L. (1986). *Blindsight: a case study and implications.* Oxford: Oxford University Press.

Weiskrantz, L. (1995). The problem of animal consciousness in relation to neuropsychology. *Behavioural Brain Research, 71,* 171–175.

Weisman, R., Hamilton, L., and Carlton, P. (1972). Increased conditioned gustatory aversion following VMH lesions in rats. *Physiology and Behavior, 9,* 801–804.

Wellman, P. J. and Boissard, C. G. (1981). Influence of fluid deprivation level on the extinction of conditioned taste aversion induced by amphetamine in female rats. *Physiological Psychology, 9,* 281–284.

Wellman, P. J., Malpas, P. B., and Wikler, K. C. (1981). Conditioned taste aversion and unconditioned suppression of water intake induced by phenylpropanolamine in rats. *Physiological Psychology, 9,* 203–207.

Wellman, P. J., Watkins, P. A., Nation, J. R., and Clark, D. (1984). Conditioned taste aversion in the adult rat induced by dietary ingestion of cadmium or cobalt. *Neurotoxicology. 5,* 81–90.

Welzl, H., Alessandri, B., and Baettig, K. (1990). The formation of a new gustatory memory trace in rats is prevented by the noncompetitive NMDA antagonist ketamine. *Psychobiology, 18,* 43–47.

White, L. (1965). Olfactory bulb projections of the rat. *Anatomical Record, 152,* 465.

Wilcoxon, H. C., Dragoin, W. B., and Kral, P. A. (1971). Illness induced aversions in rat and quail: relative salience of visual and gustatory cues. *Science, 171,* 826–828.

Willner, J., Gallagher, M., Graham, P., and Crooks, G.B. (1992). *N*-Methyl-D-aspartate antagonist D-APV selectively disrupts taste-potentiated odor aversion learning. *Behavioral Neuroscience, 106,* 315–323.

Wolfman, C., Fin, C., Dias, M., Bianchin, M., Da Silva, R. C., Schmitz, P. K., Medina, J. H., and Izquierdo, I. (1994). Intrahippocampal or intraamygdala infusion of KN62, a specific inhibitor of calcium/calmodulin-dependent protein kinase II, causes retrograde amnesia in the rat. *Behavioral and Neural Biology, 61,* 203–205.

Woods, S. C., Weisinger, R. S., and Wald, B. A. (1971). Conditioned aversions produced by subcutaneous injections of formalin in rats. *Journal of Comparative and Physiological Psychology, 77,* 410–415.

Woolf, N. J., Young, S. L., Johnson, G. V. W., and Fanselow, M. S. (1994). Pavlovian conditioning alters cortical microtubule-associated protein-2. *NeuroReport, 5,* 1045–1048.

Yajeya, R., Patino, A., Riolobos A. S., Criado, J. M., and De La Fuente, A. (1991). Passive avoidance conditioning and unitary activity in the basolateral amygdaloid nucleus of the rat. *Acta Physiologica Scandinavica, 141,* 549–553.

Yamamoto, T. (1984). Taste responses of cortical neurons. *Progress in Neurobiology, 23,* 273–315.

Yamamoto, T. (1989). Role of the cortical gustatory area in taste discrimination. In R. H. Cagan (Ed.), *Neural mechanisms in taste* (pp. 197–219). Boca Raton: CRC Press.

Yamamoto, T. (1993). Neural mechanisms of taste aversion learning. *Neuroscience Research, 16,* 181–185.

Yamamoto, T. (1994). A neural model for taste aversion learning. In K. Kurihara, N. Suzuki, and H. Ogawa (Eds.), *Olfaction and taste, XI* (pp. 471–474). Tokyo: Springer-Verlag.

Yamamoto, T. and Fujimoto, Y. (1991). Brain mechanisms of taste aversion learning in the rat. *Brain Research Bulletin, 27,* 403–406.

Yamamoto, T. and Kawamura, Y. (1972). Summated cerebral responses to taste stimuli in rat. *Physiology and Behavior, 9,* 789–793.

Yamamoto, T., Matsuo, R., and Kawamura, Y. (1980*a*). Localization of cortical gustatory area in rats and its role in taste discrimination. *Journal of Neurophysiology, 44,* 440–455.

Yamamoto, T., Yuyama, N., and Kawamura, Y. (1980*b*). Responses of cortical taste cells and chorda tympani fibers of anodal D.C. stimulation of the tongue in rats. *Experimental Brain Research, 40,* 63–70.

Yamamoto, T., Azuma, S., and Kawamura, Y. (1984). Functional relations between the cortical gustatory area and the amygdala: electrophysiological and behavioral studies in rats. *Experimental Brain Research, 56,* 23–31.

Yamamoto, T., Yuyama, N., Kato, T., and Kawamura, Y. (1985*a*). Gustatory responses of cortical neurons in rats. II. Information processing of taste quality. *Journal of Neurophysiology, 53,* 1356–1369.

Yamamoto, T., Yuyama, N., Kato, T., and Kawamura, Y. (1985*b*). Gustatory responses of cortical neurons in rats. III. Neural and behavioral measures compared. *Journal of Neurophysiology, 53,* 1370–1386.

Yamamoto, T., Matsuo, R., Kiyomitsu, Y., and Kitamura, R. (1988). Taste effects of 'umami' substances in hamsters as studied by electrophysiological and conditioned taste aversion techniques. *Brain Research, 451,* 147–162.

Yamamoto, T., Matsuo, R., Kiyomitsu, Y., and Kitamura, R. (1989). Taste responses of cortical neurons in freely ingesting rats. *Journal of Neurophysiology, 61,* 1244–1258.

Yamamoto, T., Matsuo, R., Fujimoto, Y., Fukunaga, I., Miyasaka, A., and Imoto, T. (1991). Electrophysiological and behavioral studies on the taste of umami substances in the rat. *Physiology and Behavior, 49,* 919–925.

Yamamoto, T., Shimura, T., Azuma, S., Bai, W.-Zh., Fujimoto, Y., and Wakisaka, S. (1992*a*). Induction of c-*fos*-like protein in the central nervous system of the rat following internal malaise. In R. Inoki, Y. Shigenaga, and M. Tohyama (Eds.), *Processing and inhibition of nociceptive information* (pp. 165–168). Amsterdam: Elsevier.

Yamamoto, T., Shimura, T., Sako, N., Azuma, S., Bai, W.-Zh., and Wakisaka, S. (1992*b*). c-*fos* expression in the rat brain after intraperitoneal injection of lithium chloride. *NeuroReport, 3,* 1049–1052.

Yamamoto, T., Shimura, T., Sako, N., and Bai, W.-Zh. (1993*a*). Brain sites and neural activities related to formation of conditioned taste aversion. *Neurosciences, 19,* 185–188 (in Japanese).

Yamamoto, T., Shimura, T., Sako, N., Sakai, N., Tanimizu, T., and Wakisaka, S. (1993*b*). c-*fos* expression in the parabrachial nucleus after ingestion of sodium chloride in the rat. *NeuroReport, 4,* 1223–1226.

Yamamoto, T., Shimura, T., Sakai, N., and Ozaki, N. (1994*a*). Representation of hedonics and quality of taste stimuli in the parabrachial nucleus of the rat. *Physiology and Behavior, 56,* 1197–1202.

Yamamoto, T., Shimura, T., Sako, N., Yasoshima, Y., and Sakai, N. (1994*b*). Neural substrates for conditioned taste aversion in the rat. *Behavioural Brain Research, 65,* 123–137.

Yamamoto, T., Shimura, T., Sako, N., Yasoshima, Y., and Sakai, N. (1994*c*). Some critical factors involved in formation of conditioned taste aversion to sodium chloride in rats. *Chemical Senses, 19,* 209–217.

Yamamoto, T., Fujimoto, Y., Shimura, T., and Sakai, N. (1995*a*). Conditioned taste aversion in rats with excitotoxic brain lesions. *Neuroscience Research, 22,* 31–49.

Yamamoto, T., Shimura, T., Sako, N., and Sakai, N. (1995*b*). c-*fos* immunohistochemical study on brainstem mechanisms of conditioned taste aversion in the rat. *Abstracts of the Fourth IBRO Congress at Kyoto,* p. 393.

Yasoshima, Y. and Yamamoto, T. (1997). Rat gustatory memory requires protein kinase C activity in the amygdala and cortical gustatory area. *NeuroReport, 8,* 1363–1367.

Yasoshima, Y., Shimura, T., and Yamamoto, T. (1995*a*). Single unit responses of the amygdala after conditioned taste aversion in conscious rats. *NeuroReport, 6,* 2424–2428.

Yasoshima, Y., Shimura, T., and Yamamoto, T. (1995*b*). Changes of responses in the insular cortex in freely behaving rats after acquisition of conditioned taste aversion learning. *Japanese Journal of Physiology, 45, Suppl. 1,* S188.

Yirmiya, R., Zhou, F. C., Holder, M. D., Deems, D. A., and García, J. (1988). Partial recovery of gustatory function after neural tissue transplantation to the lesioned gustatory neocortex. *Brain Research Bulletin, 20,* 619–625.

Yokel, R. A. and Ogzewalla, C. D. (1981). Effects of plant ingestion in rats determined by the conditioned taste aversion procedure. *Toxicon, 19,* 223–232.

Young, J. K. (1991). Estrogen and the etiology of anorexia nervosa. *Neuroscience and Biobehavioral Reviews, 15,* 327–331.

Yu, B. and Shinnick-Gallagher, P. (1994). Interleukin-1b inhibits synaptic transmission and induces membrane hyperpolarization in amygdala neurons. *Journal of Pharmacology and Experimental Therapy, 271,* 590–600.

Yuan, C.-S. and Barber, W. D. (1991). Parabrachial nucleus: neuronal evoked responses to gastric vagal and greater splanchnic nerve stimulation. *Brain Research Bulletin, 27,* 797–803.

Zabik, J. E. and Roache, J. B. (1983). 5-Hydroxytryptophan-induced conditioned taste aversion to ethanol in the rat. *Pharmacology, Biochemistry and Behavior, 18,* 785–790.

Zahorik, D. M. (1977). Associative and non-associative factors in learned food preferences. In L. M. Barker, M. R. Best, and M. Domjan (Eds.), *Learning mechanisms in food selection* (pp. 181–199). Waco: Baylor University Press.

Zilles, K. (1985). *The cortex of the rat. A stereotaxic atlas.* Berlin: Springer-Verlag.

Index

acetylcholinesterase 113, 117
across-neuron pattern (ANP) 80, 81
across-region pattern (ARP) 81
acquisition
  of aversive conditioned reactions 36
  circuit 46
  of conditioned immune responses 130
  day 108, 122
  of different learning tasks 117
  of memories 126
  of taste-potentiated odour aversions 36
  of the Morris water-maze task 115
  of toxiphobic conditioning 35
  process 106
  sessions 56
  trial 70, 109
adaptation 5, 14, 126
adrenocortico-tropic hormone (ACTH) 67, 68
alanine 29, 32
alcohol 63, 126
  aversion 45
  intake 22
alpha-MSH 67
amnesia 46, 55, 123
amphetamine 18, 28, 35, 64, 94, 103, 124, 125
amygdala 30, 32, 34, 42–5, 51, 55, 58, 73, 74, 83, 84, 91, 93, 98, 103, 104, 106–9, 111, 112, 127
  basolateral nucleus (BLA) 33, 34, 36–8, 72, 129
  central nucleus (CeA) 33, 35, 36, 79, 83
    corticomedial nuclei 35
  dysfunction 72
  electrophysiology of 88–91
  feedback from 55
  functional ablation of 54
  lesion 71, 73, 108
  lesion studies of 34–8
  stimulation of 35
amygdalectomy 45, 58
anatomical marker 92, 96
anaesthesia 11, 12, 47, 48, 55, 74, 101, 108, 122
  ether 11
  pentobarbital 11, 47, 48, 54
and-gate function 45
anisomycin 53, 105, 123
anodal DC current 128
anorexia 62, 63
antibiotics 123
anticipatory contrast 126
Aplysia californica 4
apomorphine 7, 23, 33, 36, 64, 69
aposematic coloration 3
area postrema (AP) 64, 79, 124, 129
  c-*fos* in 93, 94, 103
  lesion studies of 26–8
  neurons 83
  projection from 30
  stimulation of 30, 128, 129
attenuation
  of CTA 32, 54
  of neophobia (AN) 10, 14, 15, 24, 34
autonomic nervous system 35, 129
aversive response 18–21, 23, 24, 43, 51, 54, 65, 88, 89, 124
avoidance
  reaction 49, 51
  response 60, 67

barbiturates 64
basal ganglia 51
bed nucleus of stria terminalis 30, 54, 79
behavior
  active 68
  adaptive 2, 41
  aversive or avoidance 63, 67, 88, 89, 91, 98
  feeding 1, 5, 33
  ingestive 96, 101
  instrumental or operant 21, 64, 65
  licking 19
  prepared 3
  rejection 90
behavioral recovery 111–7
benzodiazepines 64
blindsight 12, 122
blocking 25
blue jays 3
bradykinin 63
brainstem 26, 58
bZIP proteins 128

c-*fos* 92, 99, 104, 105, 127
  antisense 127
  expression 85, 92–6, 99, 102, 103, 123
  immunoreactivity 94, 102–4
  immunohistochemical study 83, 84, 98
  immunohistochemical technique 124
  immunohistochemistry 81, 96

c-*fos*-like immunoreactivity (c-FLI) 58, 59, 92, 125
Ca^{2+}/calmodulin-dependent protein kinase 126
carbachol 73
cardenolide 3
catecholaminergic
  depletion 45
  mechanisms 41
chemoreceptor trigger zone 26
chemotherapy 62, 79
chemotopical segregation 81
chin rub (rubbing) 18, 65, 124
chlorpheniramine 63
cholecystokinin (CCK) 65
choline acetyltransferase (ChAT) 113, 115
cholinergic
  activity 73, 74, 113, 115
  agonists 73, 74
  antagonists 74
  blockade 115
  deafferentation 73
  mechanisms 112
  neurotransmitter system 112, 115
  release from the insular cortex 74
  stimulation 35
  transmission 116, 126
  cimetidine 63
cocaine 64, 124, 126
cognitive deficits 73, 108
comatose state 12, 48, 122
conditioned
  food aversion 4, 7
  food preference 4
  immunity 129
  immunosuppression 130
  inhibition 24
  place preference 124
  reaction (CR) 2, 36
  stimulus (CS) 60, 109, 131
  taste 58, 60, 66, 74, 130, 131
  taste aversion (CTA)
    blockade 47
    disruption 37, 54, 115
    engram 32, 46, 55, 117
    extinction of 40, 53, 58, 68
    lateralization 57
    mechanisms 26
    paradigm 14, 25, 32, 82, 92, 106, 123, 128
    plasticity 126
    retention of 19, 26, 30, 32–4, 36, 40, 41, 43, 45, 91
    retrieval of 21, 30, 39, 46, 47, 50, 53–5, 57, 58, 118–20, 128
    storage 111, 117
    to vascular taste 22
  taste preference 22, 23
conditioning
  backward 7, 69
  defense 7
  fear 89
  forward 7, 69
  higher-order 24, 70
  noise-shock 44
consciousness 12, 121
consolidation 54–6, 92, 105, 106, 117, 118, 120, 126
consummatory behavior 129
convergence 8, 30, 42, 44, 55, 70, 84
  of CS and US 101
  of taste and odour stimuli 37
copper sulfate 28
corollary discharge 5
cortical spreading depression 50, 101
corticofugal feedback 55
CREM 104, 105
crop 5, 6
CS-US
  association 2, 7, 99, 125, 127
  contiguity 121
  delay 6, 8, 22, 69
  interval 7, 11, 22, 48, 51, 69, 106, 121, 127, 129
  pairing 55, 60, 106, 128
CS-US-FB (homeostatic feedback) 71
CTA-inducing drugs 60, 63
CTA-prone and CTA-resistant rats 128
cue
  associative 69, 70
  non-gustatory 7
  odour 45, 70, 73
  taste (gustatory) 8, 29, 45
cychlophosphamide 67, 130
cycloheximide 123

D-2amino-5phosphonovalerate (D-APV) 38
decerebrate rats 43, 55, 58, 98, 99, 101
2-deoxyglucose (2DG) autoradiographic technique 105
depression 129
diarrhea 15, 66, 82, 103
dopamine 113
dopaminergic neurotransmitter system 64
dopaminergic terminals 38
drinking box 14, 17
drinkometer 19
drug abuse 122, 125

electric taste 128
electrical intracerebral stimulation 128
electrical stimulation 7, 30, 36, 55, 83, 84, 86, 96, 101, 128, 129
electroconvulsive shock 11

ELISA 130
emetic
  agents 64, 65, 124, 125
  center 125
  drugs 65, 95, 124, 125
  effects of drugs 60
  mechanisms 124
energy budget 6
entorhinal cortex (EC) 30, 40, 127
episodic memory 122
esophageal distention 94
  nerve 5
  sphincter 5
estrogen 63
ethanol 28, 64, 103
ethology 1
euphoric effects 125
evolutionary pressures 1
extinction 5, 15, 20, 23, 24, 33, 35, 37, 40, 49, 51, 53, 58, 67–70, 108, 128

fasting 10
fear 124, 125
feeding 1, 2, 5, 6, 12, 13, 23, 70
  forced 48, 51
  center 33, 91
  preference 122
fetal brain transplants 108, 116
fimbria fornix 114
flight recorder 12
floor effect 15
food
  aversion 3, 10, 62, 79, 121
  consumption 35, 40, 66
  intake 10, 35, 66, 90
  preference 4, 22, 121
foraging 6
forebrain-brain stem connections 59
forelimb flailing 43
functional
  ablation 46, 47, 54, 56, 58, 122, 126
  decortication 8, 11, 12, 50, 51, 56, 58, 101
  hemidecortication 56, 57
  lesion 53
  loop 120
  neuroanatomy 92
  recovery 108, 111–14, 116, 117, 120

gamma-amino-butyric acid (GABA) 90
gamma radiation 28
gangliosides 113
GAP-43 protein 127
gape (gaping) 18, 43, 65, 124
garden slug 3
gastrointestinal responses 103
gene targeting method 128
genetic studies 128
globus pallidus (GP) 34
glutamate 76, 78, 113
glutamic acid decarboxylase (GAD) 113, 115
grafts
  cortical 109, 111, 113–17
  functional 113
  heterotopical 116–19
  homotopical 111, 116, 118, 119
gustatory
  afferents 30, 37, 83, 101
  cortex 8, 32, 51, 53–5, 58, 105, 106, 123
  CS 4, 21, 23, 24, 28, 102, 104, 126, 127–9
  discrimination 20, 86
  long term memory (GLTM) 8, 10, 123
  neocortex 42, 45, 108, 111
  pathway 20, 46, 54, 79, 84
  preference 29
  short term memory (GSTM) 8, 129
    buffer 8
    US association 11, 12, 48, 54
  stimuli 10–12, 20, 21, 47, 51, 54, 77, 99, 100
    affective aspect of 100
    novel 14
    unpalatable 14
  thalamic nuclei 32
gustatory-visceral association 101
gustatory-visceral experience 11
gut 5, 6, 63, 65, 66, 74, 125

H7 106
head shakes (shaking) 18, 43
hedonic
  appreciation 84, 86
  aspect of gustatory information 98
  change 99
  dimension 91
  properties of food 66
  responses 43
  shift 43, 98, 103
  value 71, 88, 92, 96, 126
hedonics
  negative 96, 99, 102, 125
  of taste stimuli 88
  positive 96, 99, 102
hemidecerebrate rats 58, 104
hippocampus 30, 39, 40, 54, 73, 74, 112, 114, 127
histamine 28, 63
homeostasis 1, 2, 121
horseradish peroxidase (HRP) 44, 111
host brain 108, 111–13, 116
humoral route 83
hunger 1, 10, 35
6–hydroxydopamine 41
hypercapnia 4

hyperstriatum ventrale 105
hypothalamic-pituitary-adrenal axis 129
hypothalamus 33–6, 41, 83, 91, 93, 116, 130
  hypothalamic nuclei 26, 33
  hypothalamic lesions 130
  lateral hypothalamus 30, 35, 54, 79
  lateral hypothalamic area (LH) 30, 33
  ventromedial hypothalamic nucleus (VMH) 30, 33
hypothermia 11, 49, 101

IgG 130, 131
IgM 131
immediate early genes 92, 104
immune responses 129, 130, 131
immunohistochemical staining 92
immunological responses 130
immunosuppressive effects 131
in situ hybridization 104
insular cortex (IC) 35, 36, 41–5, 51, 58, 88, 105, 106, 113, 127
  agranular 44
  dysgranular 42
  granular 42
  gustatory 32, 105, 106
  lesions 43, 45, 108, 109, 112, 115, 118, 130, 131
insulin secretion 101
integration of gustatory and visceral signals 32, 99
internal consequences 60, 101
intracellular
  cascades 105
  events 92, 105
  messengers 126
intracranial stimulation 129
intraoral cannula 17, 19
ionizing radiation 62, 122
irradiation 28, 40, 95, 101

jun family 104

kainate 34
kainic acid 41
ketamine 47, 73
krox-24 104

labeled-line (LL) 80
lactose 18
latent inhibition 24, 40, 53, 106
lateralization 56–8
learning
  food selection 3, 5, 6, 10
  implicit 12, 121
  processes 2, 118
  recovery of 115, 116
  shock avoidance 123
  state-dependent 10, 47
  taste preference 11
  theory 121
  without awareness 122
lesions
  amygdala (amygdaloid) 37, 38, 71, 73, 108
  area postrema (AP) 28, 64, 83, 124
  brain 108, 126
    irreversible 26
  combined 33, 36, 43, 98
  cortical 44, 116
  electrolytic 29, 30, 32–4, 36, 38, 45, 108, 114
  entorhinal cortex (EC) 41
  excitotoxic 36
  functional 53, 56
  hippocampal 39, 40
  ibotenic acid 30, 33, 37, 40, 98, 125
  insular cortex (IC) 43, 45, 108, 109, 115, 130, 131
  NMDA 36, 111, 118, 130
  permanent 35, 73, 120
  preacquisition and/or postacquisition 46
  quisqualic acid 37
  reversible 36, 117, 120
  septal 73
  VPMpc 32
lick duration 19
lick-triggered rewarding stimulation 129
Limax maximus 3
limbic system 34–7, 42, 73–5, 130
lithium chloride (LiCl) 8, 15, 35, 65, 82
lobus paraolfactorius 105
long-lasting excitation 128
long-term
  memory 118, 122, 127
  potentiation 104–6, 126
lying on belly posture 103

malaise 2, 12, 19, 43, 60, 62, 66, 69, 74, 92, 101, 106, 127
mantis 3
masking solutions 121
medial forebrain bundle 125, 129
medicine effect 11, 22, 48
mediodorsal cortex 41
melanin-stimulating-hormone (MSH) 67
memory
  buffer 8
  consolidation 53
  CTA 92, 118, 127
  gustatory 53
  readout 20, 57
  retrieval 20, 34, 41
  storage 34, 41, 120

trace 56, 57
metabolic mapping 105
metaldehyde 4
methionine 4
microinjections 118, 125
microiontophoretic delivery of kainate 34
microtubule-associated protein-2 (MAP-2) 105
milkweed 3
molecular and cellular nature of CTA plasticity 126
monarch butterfly 3
morphine 64, 65, 124–6
morphological maturation 125, 126
morphological recovery 112, 117
Morris spatial water maze 115, 117
motion sickness 28
multiple bottle 15, 17
muscarinic antagonists 115, 127
muscarinic cholinergic system 127

N-methyl-D-aspartate (NMDA) 36
  antagonist 38, 72
  lesions 36, 111, 130
  receptors 38, 73, 105
nausea 2, 12, 29, 63, 66, 122, 125
neocortical grafts 113, 116, 117
neophobia 11, 12, 14, 15, 19, 24, 30, 32, 34, 45, 68, 89
neophobic response 24, 39
nerve growth factor (NGF) 113, 114
neuroimmunology 131
neuropharmacology 71
nicotine 28, 64
Novocaine 36, 38, 55
nucleus accumbens 125
nucleus basalis magnocellularis (NBM) 37
nucleus tractus solitarius (NTS) 26, 29, 79, 85, 86, 102
  caudal and intermediate portions of 103
  intermediate 94, 103, 104, 124
nutritional value 2, 4–6, 10

odour 3, 4, 7, 11, 13, 37, 38, 40, 44, 45, 69–74, 78, 79, 122
odour-taste 37, 38, 44, 45, 70–3
olfactory
  afferents 37
  bulb 32, 37, 44, 45
  bulbectomy 45
  cortex 37
oligodeoxynucleotides antisense 127
opiate users 125
opiates 125
orbital frontal neocortex (OFN) 44
oromotor responsiveness 54
orosomatic stimuli 101
overshadowing 25

palatability 14, 17, 18
paleostriatum augmentatum 105
parabrachial nucleus (PBN) 29, 54, 120
  blockade 54, 55, 57
  cls of the PBN 97
  dls 96–100
  els 96–100
  lat 30, 32
  lateral PBN 79, 83, 94, 123, 125
  lesions of PBN 30, 32 PBN
  PBNmed 30, 32, 33
passive and active shock avoidance 36
passive drip 18, 21, 43
Pavlovian conditioning 60
paw
  flail 18
  treading 124
  wipe 18
pecking suppression 3
perirhinal cortex 41, 44
permanent memory storage 111
pharmacological effects 60, 65, 126
phencyclidine 64
phosphorothioate 127
physostigmine 73
piriform cortex 34, 42, 44, 93
placing reaction 51
plastic change 47, 87, 101, 102, 126, 129
polymyxin B 56, 106
position bias 17
positive reinforcers 64, 124
post-acquisition 46
pre-acquisition 46
preference
  drinking test 67, 68
  for the location 64
prefrontal cortex (PFC) 41, 42, 113
preparedness 2, 3
priming 8, 55, 57, 101
procaine 71, 72, 86
protein kinase C (PKC) 55
  PKC inhibitors 56, 106, 127
protein synthesis 53, 105, 123
  inhibitors (PSIs) 105, 122, 126
proteosynthesis 122
psychoactive drugs 64, 124, 125
punishment 2

quinidine 3, 4
quinine 17, 43, 58, 74, 78, 81, 85, 86, 96, 99, 100, 103, 104, 124

radiation 13, 64, 122
reaction time 102

recognition 1, 20, 29, 46, 90
regurgitation 7, 10, 22, 121
reinforcement 5
restorative effects 117
retention test 15, 54, 56, 107, 129
reticular formation 29, 49
retrieval failure 25
reward 2, 125, 126
rewarding drugs 65, 124, 125
Rompun 47
rotation of the body 95

salivation 2, 35
satiation 2, 5, 6, 14
satiety center 33, 91
scopolamine 28, 64, 115, 117
seaweed 4
sensitization 24
sensory and-gate 37
serotonine 63
single bottle 14, 15, 17, 23
somatomotor cortex 41
splanchnic nerve 82, 83, 94
split brain 56
spontaneous recovery 109, 111
spout 17, 19, 20, 101, 124
spreading depression 11, 50, 53, 56, 57, 101
staurosporine 106
stress 42, 67, 81, 103, 112, 129
stress related peptides 67
substantia innominata 30
superior colliculus 58, 104
survival advantage 6
synaptic weight 8

tactile stimulation of the tongue 101
taste
  avoidance 125, 126
  basic tastes 76, 78, 80–2
  bud 20, 22, 48, 76, 78, 79
  hedonics 86, 88, 90
  interaction 69
  nerves 79, 80, 96, 128
  novel 14, 19, 24, 41, 60, 69, 74, 98, 106, 131
  quality 30, 32, 81, 82, 84–6, 99, 103
  reactivity 17–19, 24, 30, 71, 101
  receptor cells 77, 79
  responsiveness 43, 84
  shyness 124, 125
taste-potentiated odour aversions 13, 37, 38, 40, 44, 45, 70, 73, 74
taste-visceral integration 32, 84
temperature 7, 13, 48, 49, 101, 121
tetrodotoxin (TTX) 51
thalamus 30, 324, 42, 79, 83, 84, 93, 96, 106, 111, 112, 116
  medial nucleus 32
  midline and intralaminar thalamic complex (MITC) 33
  thalamic taste relay 106
  ventral 54
  ventral posterior 32, 83
  VM thalamic nuclei 112
  VPLpc 32
  VPMpc 32, 33, 79
thirst 10, 14, 17
time window 8, 11, 53, 106, 120–3
tissue grafts 108
tongue protrusion 17, 19
toxicosis 10, 35, 37, 82
toxicity index 63
toxiphobic mind system 122
transplanted neurons 108
tranquilizers 74
transplantation 108, 109, 112–14, 116
trigeminal nerve 101
trophic factor 113
tyrosine phosphorylation 74, 75, 106
two bottle 23, 66

umami 76, 78
unconditioned reflexes (UR) 121
unconditioned stimulus (US) 60, 130

vagal (vagus)
  afferent 26
  nerve 26, 28, 29, 82–4, 94
vagotomy 64, 82, 104
vestibular stimulation 28
visceral
  afferent 2, 26, 30, 35, 83, 94
  consequences 2, 3, 10, 12, 38, 40, 75, 121
  cortex 42
  sensation 26, 83, 99
  sensory modalities 83
  states 2, 13, 24, 122, 126
  stimulation 83, 84, 92, 99, 104
vitamin B 1
vomiting 2, 26, 63, 64

water deprivation 14
water intake 14, 35, 108, 129

x-rays 35

The manufacturer's authorised representative in the EU for product safety is Oxford University Press España S.A. of el Parque Empresarial San Fernando de Henares, Avenida de Castilla, 2 – 28830 Madrid (www.oup.es/en or product.safety@oup.com). OUP España S.A. also acts as importer into Spain of products made by the manufacturer.

www.ingramcontent.com/pod-product-compliance
Ingram Content Group UK Ltd.
Pitfield, Milton Keynes, MK11 3LW, UK
UKHW020923170926
475595UK00031B/37
9780198523475